Lecture Notes in Electrical Engineering 232

For further volumes:
http://www.springer.com/series/7818

Oliver Drubel

Converter Applications and their Influence on Large Electrical Machines

Springer

Author
Dr.-Ing Oliver Drubel
Siemens AG
Berlin
Germany

ISSN 1876-1100 e-ISSN 1876-1119
ISBN 978-3-642-36281-1 e-ISBN 978-3-642-36282-8
DOI 10.1007/978-3-642-36282-8
Springer Heidelberg New York Dordrecht London

Library of Congress Control Number: 2012956231

© Springer-Verlag Berlin Heidelberg 2013

The content of this book is identical with the habilitation of the author. The work has been accepted by the Technische Universität Dresden with the disputation on 21th of Nov. 2011.

This work is subject to copyright. All rights are reserved by the Publisher, whether the whole or part of the material is concerned, specifically the rights of translation, reprinting, reuse of illustrations, recitation, broadcasting, reproduction on microfilms or in any other physical way, and transmission or information storage and retrieval, electronic adaptation, computer software, or by similar or dissimilar methodology now known or hereafter developed. Exempted from this legal reservation are brief excerpts in connection with reviews or scholarly analysis or material supplied specifically for the purpose of being entered and executed on a computer system, for exclusive use by the purchaser of the work. Duplication of this publication or parts thereof is permitted only under the provisions of the Copyright Law of the Publisher's location, in its current version, and permission for use must always be obtained from Springer. Permissions for use may be obtained through RightsLink at the Copyright Clearance Center. Violations are liable to prosecution under the respective Copyright Law.

The use of general descriptive names, registered names, trademarks, service marks, etc. in this publication does not imply, even in the absence of a specific statement, that such names are exempt from the relevant protective laws and regulations and therefore free for general use.

While the advice and information in this book are believed to be true and accurate at the date of publication, neither the authors nor the editors nor the publisher can accept any legal responsibility for any errors or omissions that may be made. The publisher makes no warranty, express or implied, with respect to the material contained herein.

Printed on acid-free paper

Springer is part of Springer Science+Business Media (www.springer.com)

... do not forget a trusted friend

To my parents
Marlis und Manfred Drubel

Overview

Converter driven applications are applied in more and more processes. Almost any installed wind-farm, ship drives, steel mills, several boiler feed water pumps, extruder and many other applications operate much more efficient and economic in case of variable speed solutions. The boundary conditions for a motor will change, if it is supplied by a converter. An electrical machine, which is operated by a converter, can no longer be regarded as an independent component, but is embedded in a system consisting of converter and motor.

An overview of existing converter designs for large electrical machines will be given and methods for the appropriate calculation of machine phenomena, which are implied by converters, will be derived. In the power range above 500 kVA a wide variety of converter types exists. Either diode rectifiers as well as voltage source converters are dominant converter types in several applications. Additionally cyclo-converters and other current source inverters are still applied in some niche markets. The converter design is different for the individual converter types as well. Voltage source inverters are built f.i. as two-, three- or multi-level converters. The explanation of the individual phenomena is analysed either on machine converter systems, where a phenomena is especially distinct, or for the most dominant converter machine combinations within the market. In wide industrial segments the asynchronous machine supplied by a voltage source converter is this dominant system. Nevertheless shaft voltages can be excellently investigated on large turbo-generators with load commutated current source inverters (LCI).

Due to the converter inherent higher voltage harmonics and pulse frequencies special phenomena are caused inside the machine. These phenomena can be the reason for malfunction. Additional losses create additional temperature increases or voltage peaks mean higher stress levels for the insulation. Torque ripple can occur, which endanger the mechanical shaft system and last but not least shaft voltages are induced, which are sometimes sufficient in amplitude to damage bearings or to disturb sensors of the protection arrangements.

Variable speed means also, that effects, which would be related to the grid frequency for direct on line applications, will occur for a variable fundamental frequency. Electromagnetic forces, which act as source for noise, will be activated with variable frequencies. Eigen-modes, which have not been excited from the grid frequency, may be dominant due to resonance effects at converter imposed speeds. Direct on line machines can be operated at low damped over-critical speeds, if the distance to the closest critical speed is for instance 15 %-20 % of the rated speed. In a variable speed application this constraint is not sufficient. Low damped critical speeds should not occur within the complete speed control range and 15 %-20 % beside of it.

Last but not least two typical converter applications will be analyzed, which are typical for most of the converter applications. Both have certain relevance for the global warming. Converter pump drives will contribute towards energy efficiency in a system and the consideration of efficiencies in steel mill applications means a real challenge regarding their overload requirements.

Most chapters are based on the authors' industrial experience in combination with available literature. Deep investigations in insulation systems are important to design a machine, which is appropriate for converter applications. Due to the authors' lack of experience in this area, chapter 8 is based strongly on the work from Dr. Kaufhold, [8.5].

The work on this book lasted several years. Several people supported me during this time in different ways. Prof. S. Kulig motivated me to start at all. He created the awareness of a scientific harbor in Dortmund, where it is possible to get support and new ideas at any time. Especially Dr. S. Exnowski led me through the labyrinth of numeric structure dynamic calculations for the chapter of noise. The final version of the thesis is the result of the dedicated encouragement from Prof. W. Hofmann in reviewing and discussion on individual topics. Prof. K. Reichert and Prof. A. Lindemann took over the second review. I am very conscious of the support, which I received from several colleagues in Birr and Nürnberg. I want especially to thank K. Bauer for the voltage measurements, as well as H. Kleinod and Dr. C. Mundo for the good will, with which my scientific "hobby" was looked at. Last but not least I am in debt to my family Perrine, Lara and Pierre for their patience during the last years.

Herscheid, July 2008   Oliver Drubel

Contents

1 Introduction 1
    1.1 Industrial Requirements for Variable Speed Drives and Applications
        of Brushless Exciters 1
    1.2 Research Activities in the Field of Large and Medium Size Converter
        Drives... 5

2 Typical Converter Designs for Electrical Machines 9
    2.1 Overview of Converters and Their Application................................. 9
    2.2 Naturally Commutating Converters 10
    2.3 Forced Commutating Converters... 16
    2.4 Harmonic Content in the Voltage and Current Characteristics
        of Different Converter Types... 22

3 Calculation-Methods for Converter Fed Electrical Machines 27
    3.1 Analytical Calculation Methods for Special Operation Points
        and Machine Model Integration into Circuits................................. 27
    3.2 Non-linear Transient Time-Stepping Numerical Field Calculation
        with Integrated Circuit Elements 35

4 Additional Losses Due to Higher Voltage Harmonics 45
    4.1 Overview of Converter Dependent Loss Components...................... 45
    4.2 Eddy Current Distribution within the Stator and Rotor Winding 46
    4.3 Additional Iron Losses within the Stator and Rotor Lamination 50
    4.4 Eddy Current Losses within Massive Magnetic Material 52
    4.5 Loss Measurement within Magnetic Material 63
    4.6 Additional Losses for Different Voltage Source Converter Types........... 65

5 Converter Caused Torque Oscillations... 69
    5.1 Interaction between Stator Field Components 69
    5.2 Calculation of Electromagnetic Torque Oscillations 71
    5.3 Torsional Shaft Oscillations and Design Rules................................ 77

**6 Noise Based on Electromagnetic Sources in Case of Converter
   Operation**.. 85
    6.1 General Overview of Noise Calculation 85
    6.2 Determination of the Electromagnetic Force Modes and Amplitudes 86

X   Contents

6.3   Eigen-Modes and Mechanical Calculation Methods 95
6.4   Influence of Pulse Frequency and Converter Operation on Noise
Phenomena...101

7   Converter Caused Shaft Voltages 109
7.1   Overview of Different Shaft Voltage Types...109
7.2   Circumferential Flux and Capacitive Imposed Shaft Voltages...............110
7.3   Measurements of Current Path and Voltage Transients.........................119
7.4   Shaft Grounding and Converter Concepts 129

8   Insulation Strategies in Converter Driven Machines 135
8.1   Overview of Converter Implied Insulation Stress..................................135
8.2   Potential Distribution within a Converter Supplied Electrical
Machine 136
8.3   Voltage Peaks due to Wave Reflection...137
8.4   Influence of the Voltage Characteristic on the Insulation Life Time......141
8.5   Influence of the Converter Type on the Insulation System....................149

**9   Converter Applications in Typical Examples of Energy Efficient Pump
Systems and in Processes with a Strong Overload Characteristic............153**
9.1   Energy Saving Potential for Boiler Feed Water Pumps in Part Load
Operated Coal Power Plants for Different Drive Arrangements..............153
9.2   Influence of Speed and Overload Capability of Steel Mill Drives
on the Motor Efficiency 158

10   Conclusion 161

Attachment 165
A.1   Parameters of Asynchronous Machines...165
A.2   Parameters of Synchronous Machines 168

Symbols..169

References 181

Index 189

Chapter 1
Introduction

1.1 Industrial Requirements for Variable Speed Drives and Applications of Brushless Exciters

Electrical machines in the power range between 500 KVA and 2000 MVA are used in combination with power electronics in several industry segments. They can be split in the following main areas: Energy, steel and metal, paper, oil and gas, chemical, mining, food and transport. The individual manufacturing processes cause special speed and performance requirements for the drives. A wide variety of converter-machine solutions is needed to cope with these requirements. An overview of different converter applications is given in fig. 1.1.

The energy industry uses a wide range of electrical machines together with power electronics. Within conventional combined cycle power plants turbo-generators are converter fed to start the gas turbine. These converters reach sizes up to 14 MW [1.1, 1.2]. They are current source converters with d.c. link chokes. Further power electronic elements are used for the supply of the excitation of turbo-generators up to 2000 MVA. Either they are fed over slip rings or they are fed over brushless exciters feeding rotating diodes. The power of these devices may reach up to 10 MW. The second largest variable speed drive in steam power plants is given by the boiler feed water pump. Several coal fired power plants are used for the control of the electrical energy generation in order to compensate the strong variability of the contribution from wind farms within the German electricity grid. The feed water flow needs to be adjusted for part load operation. This can either be done by orifices or by the rotation speed of the pump. In general a part of the motor power in line with the slip is converted into losses. Energy efficient part load operation is only possible in case of a converter driven machine or if one of several pumps is switched of completely. Therefore boiler feed water pumps are more and more driven by converter fed drives. Their speed requirements need either a stiff motor design for a speed range from 1000 rpm-3000 rpm or high speed drives up to 5000 rpm or 6000 rpm. Whereas steam turbine plants start the turbine-generator shaft by the steam turbine and large gas turbine-generator arrangements are started by the converter fed generator, smaller gas turbines are often started by pony motors. These pony motors are used to start up gas turbines up to their ignition speed. They consist often of transient operated drives up to speeds of 5000 rpm or higher. Special requirements for their short time overload characteristic will be essential for a cost effective solution. Beyond these conventional power

O. Drubel: *Converter Appl. & their Influence on Large Electr. Mach.*, LNEE 232, pp. 1–8.
DOI: 10.1007/978-3-642-36282-8_1   © Springer-Verlag Berlin Heidelberg 2013

1 Introduction

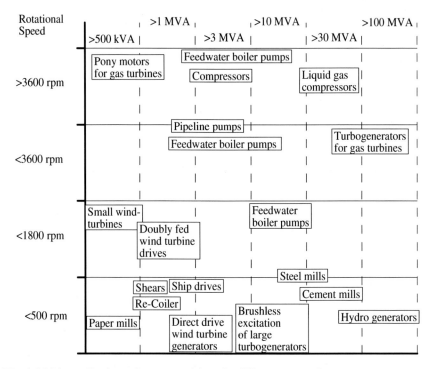

Fig. 1.1 Main applications of converter drives for different power classes

plant applications the relatively new energy segment with wind turbines reaches in the meantime 15-20 GW newly installed plant capacity per year. The wind turbine applications in the range between 2-5 MW are always variable speed drives. Five main technical directions are widely used:

- Double fed induction machines, coupled over a gear box to the turbine,
- Converter fed permanent magnet generators, coupled over a gear box,
- Converter fed squirrel cage induction generators, coupled over a gear box,
- Converter fed multi-pole-synchronous generators, directly coupled to the turbine,
- Converter fed multi-pole permanent magnet generators, directly coupled to the turbine.

Whereas the converter of the double fed induction machine needs to be designed only for 30%-50% of the turbine power the converters of the synchronous machine approaches have to be designed for the full turbine power. All five drive designs exist in parallel. The variety of drives is especially in the power industry relatively wide. An overview of energy specific drive applications is given in table 1.1.

Table 1.1 Energy specific drive applications

Application	Type of converter	Rating (kW)	Converter speciality	Speciality electrical machine
Large gas turbine start up	Current source converter	14000	Solutions with one d.c. choke only may cause high shaft voltage peaks	Low laminar cooling at low speed; Shaft grounding system; Losses in Roebel bars during high frequency currents
Static excitation of turbo-generators	Thyristor rectifier	10000	Limit in voltage peaks due to slip rings	Shaft grounding principles
Rotating excitation	Diode rectifier	10000	n-1 principle in diode arrangement	Design of rotating diodes; Circulating currents in delta- or polygon connected windings
Feed-water Boiler pump	Voltage source converter	5000	Variable speed range from 1500 to 3600	No Eigen-frequencies within a wide speed range up to 3000 rpm or 3600 rpm
Wind turbine generator	Voltage source converter	5000		High torque machines
Wind turbine generator	Doubly fed induction machine with voltage source converter	2500	Limit in voltage peaks due to slip rings; Speed control in a range from 100% to 70% n_n	Special requirements for shaft grounding. Special mechanical design has to cope with the needs of a wind turbine application
Wind turbine generator	Voltage source converter for permanent magnet machines	5000	Special converter protection necessary due to voltage induction of permanent magnets	Permanent magnet generator is coupled over a gear box
Pony motor	Voltage source converter	2000	Special transient capacity	High motor speeds; Strong transient overloads

Another area of low and medium voltage drive applications is given within the steel industry. The variation in the application is relatively small within this industry. The requirements to the machine dynamic and its control quality are much higher than in energy. Strong overload requirements up to 300 % of nominal load are coming together with control areas of constant power, which can be up to four times the speed range of constant torque. The rolled steel is transported over a transportation lane towards the roll stand. Here it is pulled between the rolls causing a jump in the torque requirement of the drive. The rolled steel is elongated forward and backward until its thickness is such that it may be coiled. The coiler and de-coiler are torque and speed controlled. An overview of the main drives in a steel mill is given in table 1.2.

Table 1.2 Main converter fed drives in a steel mill

Application	Type of converter	Rating	Converter speciality	Speciality electrical machine
Coiler	Voltage source	1.2 MW	Special control requirements	Large stalling torques
Reversing roughing stand	Current source	12 MW	Cost effective solution due to thyristor elements	Large stalling torques
Reversing edging drive	Voltage source	1.5 MW	Special control requirements	Large stalling torques
Shear drive	Voltage source	1.5 MW	Special control requirements	Large stalling torques; low inertia moment
Finishing stand	Current source	12 MW		
Bar and wire rod mills	Voltage source	8 MW	Special control requirements	Large stalling torques
Descaling pumps	Voltage source	3 MW	No special requirements	Pump characteristic

A further area of drive applications is given by fluid flow engines. These applications can be found within the oil and gas industry as well as within the chemical and food industry. Fluid flow engines of any kind are driven with variable speed drives. The main motivation for the speed variation is given by the gain in efficiency of the process, if the pump or compressor can be speed controlled. Direct on line drives would be controlled in flow by hydraulic orifices, which cause additional pressure loss. Table 1.3 shows different fields of converter applications.

Table 1.3 Converter fed compressors and pumps

Application	Type of converter	Rating	Remarks
Speed control of compressors	Current source	65 MW	Mechanical design; over-critical operation; Low leakage reactance for commutation
Speed control of compressors	Voltage source	3 MW	Mechanical design due to high rotation speeds, over-critical operation
Speed control of pumps	Voltage source	6 MW	Stiff mechanical design for motor operation up to 3600 rpm
Ship propulsion	Voltage source	5-30 MW	Motor is directly included in the propeller

Even though requirements for converter drives have been shown for a variety of processes, some have not been described explicitly. In most of the cases analogies to the explained processes may be found. Additionally only main process requirements are given. Detailed drive specifications regarding harmonic disturbances, drive and motor protection, necessary standardized tests or general machine specification can be found in [1.3, 1.4, 1.5, 1.6, 1.7]. Those converter caused machine phenomena, which endanger a proper operation in line with the basic process requirements, will be dealt within this work in detail. Especially shaft voltage, torque ripple, additional losses, noise and insulation stress are areas, where the converter supplied drive has to be handled differently to a sinusoidal supplied fix speed drive. These areas are again and again subject of research and development activities [1.8, 1.9, 1.10].

1.2 Research Activities in the Field of Large and Medium Size Converter Drives

The main focus of research activities in electrical machines and converters is in most of the cases the converter or motor itself. Unwanted phenomena within the motor are mitigated within a second step. Converters are available from 500kVA up to 100 MVA and more. These drive powers are realized by different machine-converter systems, see fig. 1.2, which are often especially in the larger power range optimized to the individual application. Converter development is mainly driven by costs and in the oil, gas and energy industry by reliability and efficiency. Main cost drivers are semi-conductive elements. Therefore one main focus of development in the power range between 500 kVA several 100 MVA is the optimization of semi-conductive elements utilization for instance by special switch algorithm. The pulse muster is often optimized by simulation under constrains, which are given by the acceptable harmonic content of the terminal and grid voltage. Beside the software itself, development efforts target to utilize simpler power electronic elements or converter topologies [1.11]. Main development tools are based upon circuit models with concentrated elements. Electrical machines are modeled with equivalent models. These models are in most of the cases sufficient. In [1.12] it is shown that even strong modifications within the damper structure of electrical machines can often not be detected at the terminals. More detailed mathematical machine models are necessary, when the influence of the converter on the electrical machine has to be investigated. Even though electrical machines are built since more than 100 years [1.13, 1.14] their application together with frequency converters causes effects, which had not been seen before. Special challenge is given to frequency phenomena within several 100 kHz, through which special attention to capacitive effects is drawn.

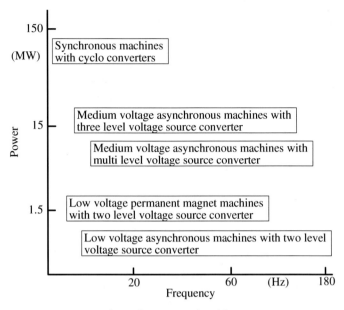

Fig. 1.2 Converter motor systems depending on speed and frequency

A more detailed machine investigation than with concentrated elements is often necessary in some converter caused phenomena like losses, shaft voltages or noise. Whereas developed methods and programs enable the skilled engineer already to solve nearly all field problems in electrical machines, which occur in reality, some gaps exist still. In the development of wind turbine generators and converter driven industrial machines numerical tools are used either in two-dimensional versions or from time to time in three-dimensional ones. Today's main application is damage investigation, mechanics and hot spot investigation. They are used to confirm an analytically determined design or to investigate ideas beyond existing design rules. They are still not powerful enough to substitute experience and analytical methods in damage prediction or for design calculations and design optimization. A schematic way to develop in these directions is given within fig. 1.3. Three main directions are emerging out of numerical field calculations. One will be the further development of the coupling between numerical and circuit models. This approach is especially necessary to investigate converter imposed phenomena in electrical machines more in detail. The second one is pure numerical, more and more integrating electrical, thermal and mechanical problems. The third main focus is given by the user interface. The development of virtual reality could open a kind of revolution in the future machine design process.

A special driver for the future in electrical machines can be seen within the area of wind energy. Especially high torque motors of very low speed are a challenge. The speed goes down to about 10 rpm. A 5 MW wind turbine generator has to provide the same torque as a 1500 MW turbo-generator of a large steam power plant. Permanent magnet generators are used for instance.

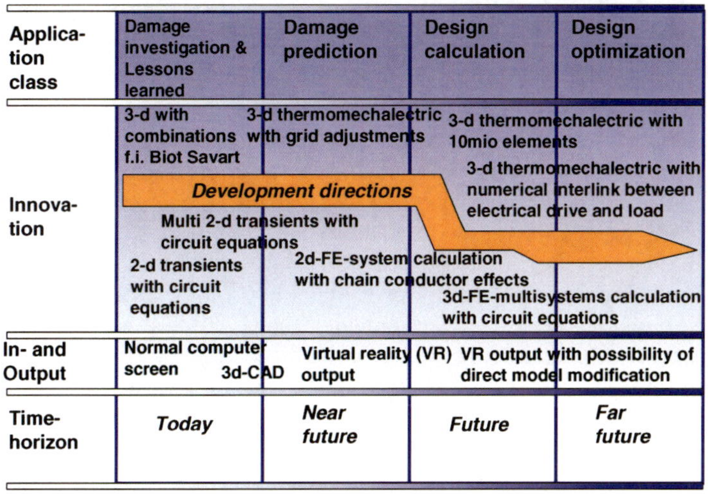

Applica-tion class	Damage investigation & Lessons learned	Damage prediction	Design calculation	Design optimization
Innova-tion	3-d with combinations f.i. Biot Savart	3-d thermomechalectric with grid adjustments	3-d thermomechalectric with 10mio elements	
	Development directions		3-d thermomechalectric with numerical interlink between electrical drive and load	
	Multi 2-d transients with circuit equations			
	2-d transients with circuit equations	2d-FE-system calculation with chain conductor effects		
			3d-FE-multisystems calculation with circuit equations	
In- and Output	Normal computer screen 3d-CAD	Virtual reality (VR) output	VR output with possibility of direct model modification	
Time-horizon	*Today*	*Near future*	*Future*	*Far future*

Fig. 1.3 Development directions in numerical tools [1.15]

The next step for higher utilizations could be three-dimensional structures. Tools will be needed, which do not only allow to calculate once such a structure, but which enable the designer to investigate within two weeks 50 variants with up to 10 mio. elements. The other design variant for wind turbine generators is often double fed at f.i. 1500-1800 rpm. The wind turbine is coupled via a gear box to an asynchronous slip ring motor. Due to the hard switching of IGBTs a capacitive current flow perpendicular to the conductors will occur in the rotor. This field problem of chain conductors is so far approached with concentrated circuit elements. The challenge is given by the right choice of parameters. It needs to be a transient field calculation, which will handle the change in machine parameters inherently allowing not only for axial current flow, but for the perpendicular capacitive component as well. It should be possible to integrate the simulation of the control and the converter in a grid of circuit elements together with 2-d and 3-d FE elements to see the influence of transients in 3-d structures keeping the short solution time of circuit elements. Temperature calculations within semiconductors during overload could be integrated in this approach as 3d-FE structures as well. Beside the development of each component by itself main challenges result through the interaction of the electrical machines with converters. The circuit driven converter design interacts more and more with the machine design and local 3-dimensional problems. A 3-d transient system, which is capable to calculate f.i. also noise in interaction with the converter are often beyond calculation capacities. Several investigations deal mainly with the electrical machine, which is specially designed for converter application or shows special effects due to converter operation.

Even though numerical field calculations of the overall system may allow a deeper understanding for the individual converter caused phenomena in electrical machines, measurements and calculation methods exist, which allow already for a proper description and pre-calculation of some phenomena. Actual research results are found in the area of converter imposed shaft voltages, additional losses as well as in the area of noise and converter adapted insulation systems [1.8, 1.9, 1.10].

The following chapters combine principle research results, which show the influence of converter applications on large electrical machines, with industrial experience and measurements in realized machines and drives. Based on an overview of typical converter designs in chapter two and on different calculation approaches for induction- and synchronous machines in chapter three, individual effects are analyzed in the following chapters.

Main effects in electrical machines due to converter operation are given by additional losses, converter caused torque oscillations, noise based on electromagnetic forces, shaft voltages or the implied insulation stress. Losses, torque and electromagnetic forces are caused by harmonic current and flux content. The calculation of converter caused shaft voltages and insulation stress impose the consideration of capacitive phenomena as well. The chapters are structured accordingly. In chapter four additional losses are investigated. Beside loss calculation due to higher harmonics within the stator and rotor winding, additional losses in laminated and massive magnetic material are determined. In chapter five the implication of converters on torque oscillations is shown. The calculation procedure is developed on a fourier harmonic basis. The torque is always a result of the force integral around the air-gap. Noise phenomena are a result of local force distributions within the air-gap. They are analyzed in chapter six based on numerical methods. Especially analytical methods, which are applied to cylindrical shells, are within large machines not sufficient to consider the stiffness of the form wound coils, nor the fact, that force modes can excite mechanical modes of different order due to machine asymmetries. In chapter seven shaft voltages in large electrical machines are presented. The chapter is split in three parts. In a first part the phenomena is described on a theoretical base, in a second part measurements are shown and last but not least shaft grounding concepts as well as rules are given for different machine arrangements. Similar to shaft voltage implications the insulation stress due to converter operation is mainly caused due to the voltage surges during the electronic switching. Chapter eight describes several effects in machines with randomly wound and form wound coils. Whereas the other chapters are within the authors own experience, the converter effects on insulation stress is based strongly on literature, [8.5, 8.6].

After the detailed analysis of the converter imposed influence on electrical machines in chapter four to eight, chapter nine describes two typical converter applications. A pump operation as typical example of a fluid machine and a drive for a steel mill as example for strong overload requirements is chosen.

Chapter 2
Typical Converter Designs for Electrical Machines

2.1 Overview of Converters and Their Application

Converters are applied within a broad field of industrial processes. The actual converter design is determined by several boundary conditions of the individual application. Most important are costs and reliability in comparison with alternative solutions. A constraint is often the maximum power output, which can be realized with existing semi-conductor devices. Especially applications with high level power performance request excellent efficiencies as well. A wide variety of converter types is the result of the difference in those boundary conditions. An overview of main types and their applications is given in table 2.1.

Table 2.1 Converter types and their applications

Converter type	Application	Speciality	Remarks
Two-level voltage source inverter bridge (VSI)	Cranes, steel mills, ships, pumps in low voltage design	Three phase bridge	Systems are highly standardized
Clamped multi-level VSI	Challenging drive dynamics like steel mills	Voltage levels defined by diode clamped capacitive elements	Mainly realized with three-level designs
Cascaded-Multi-level VSI	Medium voltage pumps and compressors	Several low voltage elements are in series	Redundant current paths can be integrated
Multi-phase diode bridge	Brushless excitation	Rotating diodes	Control by thyristors in the stator
Current source inverter	Start up of gas turbines	d.c. link with current choke	Load commutated
Cyclo-converter	Cement mills	Direct thyristor converter	Load frequency limited to ca. 40% of grid frequency

Table 2.1 is limited to the most dominant converter types in a power range above 250 kW up to 100 MW for drives. Dominant converter types are based on voltage source inverter bridges. The design can be found in all important

industries like oil & gas, energy, chemical, cranes, ships, metal. Special applications like low speed drives in cement mills fit well to cyclo-converters. A wide area of applications for rotating diodes is given by brushless synchronous machines. They are used in small synchronous machines of some few MW as well as in large turbo-generators up to more than 1000 MW. The rated power of the rotating diode brushless exciter itself may reach already 10MW on its own. Converters for high voltage d.c. links, metal melts, d.c. machines or special drives for transportation are not considered explicitly. They are in general based on the same principles or rectifier parts than those of table 2.1. Harmonics in the power supply of high voltage d.c. links and metal melts imply in special cases high stress levels to turbo generator shaft lines which have to be investigated in those individual cases. Several publications deal with this phenomenon [2.1, 2.2]. The calculation principles are in line with those, which are given in chapter 5 for voltage source inverter bridge type converters. Specialties in transportation are often based on the electricity supply of the railway system. The supply depends on the history of individual countries and has to fit to those circumstances.

Independent upon the converter type a general definition of circuits can be given [2.3]:

- A rectifier tends always to derive d.c. power from an a.c. source.
- An inverter converts a.c. power from a d.c. source.
- The term converter is generally used for both types of circuits or apparatus and also for apparatus, which consist of the combination of rectifier and inverters.

Beside the classification, which is based on the main converter purpose, converters are often differentiated in two classes based on their operation principles. Naturally commutated converters are commutated by some external voltage sources from the grid or load. Forced commutated converters are supplied with semi-conductive devices, which can commutate the current flow by themselves. This kind of classification is quite dominant in literature [2.3, 2.4, 2.5, 2.6, 2.7]. Beside the two families, three elementary types of semi-conductive circuits can be distinguished. Converters with a d.c. link are of a bridge type. Direct connections between the load and the supply describe a kind of matrix. The third kind of circuit is based on applications with loads in the neutral branch of the converter transformer.

2.2 Naturally Commutating Converters

Naturally commutating converters are based on semi-conductive devices like diodes or thyristors, which can not be switched off inherently. The device is switched off, if the current flow is reversed by external voltages. Simplest converter designs, which use the supply voltage for the current commutation are given by single way rectifiers, fig. 2.1.

2.2 Naturally Commutating Converters

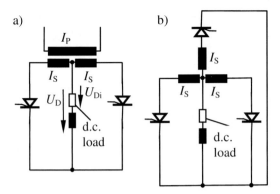

Fig. 2.1 a) Bi- and b) three phase single-way rectifiers [2.3]

Single way rectifiers do rectify the terminal voltage towards a neutral point. They cut off one half wave of the supply. In case 2.1 a) a d.c. voltage U_D and a d.c. current I_D will occur. The bi-phase transformer has to be rated with the following apparent power [2.7]:

$$S_{Tr} = \frac{1}{2}(I_p U_p + 2 \cdot I_s U_s) = \frac{(1+\sqrt{2})\pi}{4\sqrt{2}} I_D U_{Di} = 1.34 \cdot I_D U_{Di} \qquad (2.1)$$

In case 2.1 b) similar results are given by:

$$S_{Tr} = \frac{1}{2}(3 \cdot I_p U_p + 3 \cdot I_s U_s) = \frac{\left(1+\sqrt{\frac{2}{3}}\right)\pi}{3\sqrt{2}} I_D U_{Di} = 1.35 \cdot I_D U_{Di} \qquad (2.2)$$

The utilization of the converter transformer is poor in comparison with a bridge converter, but only half the number of diodes or thyristors is needed. Even though the semi-conductive elements need to be rated for twice the voltage of a bridge rectifier, single way rectifiers are used in processes, which require large currents at low voltages.

Bridge rectifiers are rectifiers, where the grid or generator terminal lead is connected to a bridge, which is built out of two semi-conductive devices. A typical bridge arrangement for a brushless exciter is shown in fig. 2.2.

The current commutation in the diodes is forced by the induced voltage in the brushless exciter. Each individual winding of the brushless exciter will lead a positive and negative current flow. Even though the current characteristic is not sinusoidal, the utilization is higher. A three-phase bridge rectifier would need only a transformer apparent power acc. equation (2.3):

$$S_{Tr} = \sqrt{3} \cdot I_s U_s = \frac{\pi}{3} I_D U_{Di} = 1.05 \cdot I_D U_{Di} \qquad (2.3)$$

The comparison between equations (2.1) – (2.3) is always based on ideal d.c. currents within the load and block wise currents in the transformer winding. In reality this is not the case. Fig. 2.3 shows the current course in the armature winding of a brushless exciter. The exciter-winding is polygonal connected. The current characteristic is not based on current blocks at all.

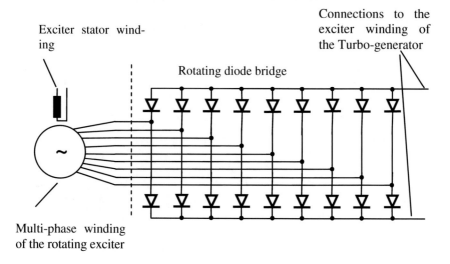

Fig. 2.2 Polygonal brushless excitation system with rotating diodes

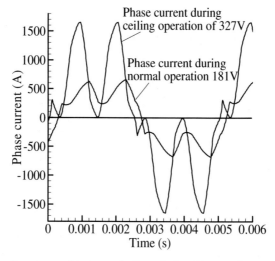

Fig. 2.3 Current flow in a multi-phase winding of a brushless exciter

Naturally commutated semi-conductive devices can be connected as rectifier. Here they are in general commutated by the supply voltage. If they are combined in an inverter circuit the reactive power for the commutation has to be generated by the load. Asynchronous machines can not provide the required commutation

2.2 Naturally Commutating Converters

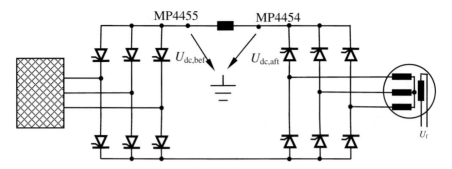

Fig. 2.4 Load commutated current source inverter

voltage inherently. Synchronous motors can be operated on naturally commutating inverters. An example is shown in fig. 2.4.

The example in fig. 2.4 exists out of two six-pulse bridges with a choke in the d.c. link. The choke needs to smooth the current in the d.c. link. Fig. 2.5 shows a measurement of the voltages against ground in the d.c. link before and fig. 2.6

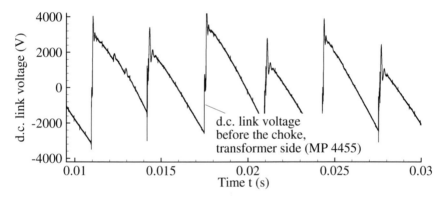

Fig. 2.5 Measured d.c link voltage $U_{dc,bef}$ before the choke

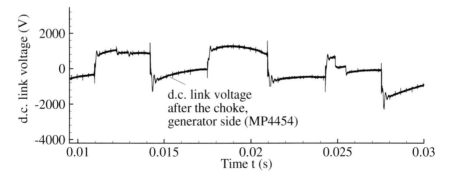

Fig. 2.6 Measured d.c link voltage $U_{dc,aft}$ after the choke

after the choke. Two effects can be seen. The ground potential is floating and the choke reduces the voltage peaks and pulsation towards the inverter side strongly.

The converter design of an asymmetrical d.c. link with only one choke is quite cost effective, but does not provide the damping effect in the other d.c. link. Frequency converters of the current source type are applied for instance for the start of larger gas turbines. The converter for a typical 250 MW gas turbine may reach ratings up to 14 MW. The converter is loaded directly by the turbo-generator. In order to reduce shaft-voltages, which are caused in especially high amplitudes by voltage surges of the non damped d.c. link, current source converters with symmetrical d.c. links are applied.

Another kind of naturally commutating converters is defined by direct converters. A famous version of a direct converter is the cyclo-converter. Naturally commutating cyclo-converters are used in a frequency range below 50% of the supply frequency. Whereas in principle three-pulse converters, see fig. 2.7, are sufficient, for lower frequencies, the frequency range up to 50% of the supply frequency can be used by six-pulse frequency converters. The drive of a three-pulse

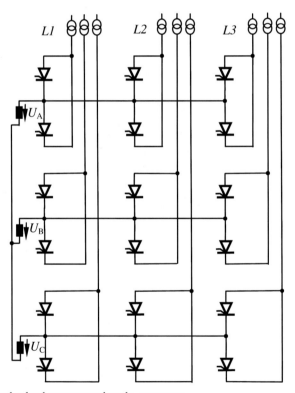

Fig. 2.7 Three-pulse load commutated cyclo-converter

2.2 Naturally Commutating Converters

converter with non insulated loads needs during start conditions an additional cable link towards the neutral of the supply transformer until all load currents sum up to zero, [2.7]. This is not shown in fig. 2.7. Additionally each bridge is supplied galvanic separated. Each phase of the load can be connected to the line voltage during the positive time period and during the negative time period.

A six-pulse cyclo-converter is based on six six-pulse bridges. Each phase can be positively or negatively connected to the load independently upon the phase of the line voltage. They operate either on galvanic separated motor windings or they must be supplied by different transformer windings. An example for a six-pulse cyclo-converter is given in fig. 2.8.

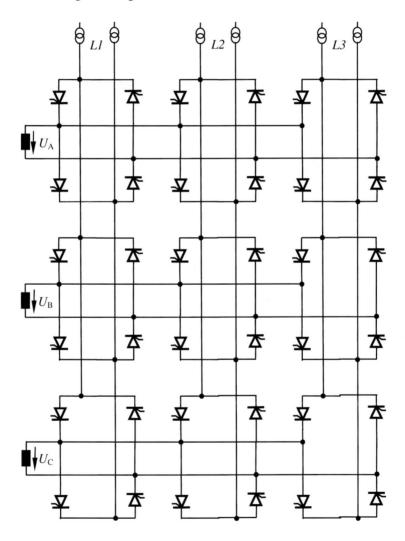

Fig. 2.8 Six-pulse load commutated cyclo-converter with insulated load

Each thyristor is commutated in that way, that the load voltage U_A is modulated by the delay angle α in time, see equation 2.4, [2.8]:

$$U_{A,\text{eff}} = \frac{3}{\pi}\sqrt{2}\, U_{RS} \cdot \cos\alpha(t) \tag{2.4}$$

The motor frequency is defined by the modulation of the delay angle. Each thyristor can be fired only once per period of the supply. This is the main reason, that no terminal voltages with base frequencies above 50 % of the supply frequency can be achieved in reasonable quality.

2.3 Forced Commutating Converters

Converter types with forced commutation are in the upper power range for most of the applications of voltage source bridge types. A basic two-level bridge type converter circuit is shown in fig. 2.9.

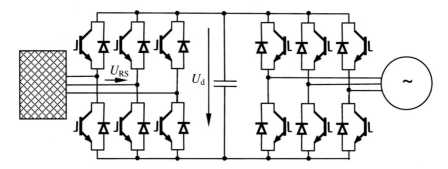

Fig. 2.9 Two-level voltage source converter with six-pulse bridge

The converter consists of twelve semi-conductive devices like IGBTs, insulated gate bipolar transistors. They built one rectifier- and one inverter-bridge. Both bridges are connected via a d.c. link. The voltage pulsation of the d.c. link is smoothened by a capacitor. The ideal d.c. link voltage $U_{d,\text{mean}}$ is given by equation 2.5:

$$U_{d,\text{mean}} = \frac{3}{\pi}\sqrt{2}\, U_{RS} \tag{2.5}$$

This voltage level can be connected to the motor terminals. A common modulation of the voltage blocks is the pulse-width modulation method. Higher voltage harmonics are reduced by an optimized voltage block width. Due to the voltage surges of the individual block switching with very small time constants voltage waves occur on the cable between the converter and the motor. These waves are reflected at the terminals as well as within the motor strands. Whereas these peaks are important for some phenomena like the stress in the insulation

2.3 Forced Commutating Converters

system the imposed machine flux is important for other converter related phenomena like machine losses. A typical voltage characteristic has been measured between motor terminals and a neutral point. The result is given in fig. 2.10 together with the flux characteristic.

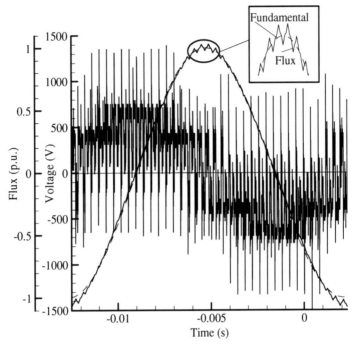

Fig. 2.10 Measured terminal voltage and flux characteristic of a 690 V motor, which is fed by a two-level converter

Two-level converters are often used in low voltage applications. Costs dominate the design and the converter control. The application of relatively small pulse frequencies allow for a larger utilization of the semi-conductive elements. Losses during the switching phase contribute with about 50% of the overall losses in the devices. The low pulse frequency does unfortunately impose higher harmonics on the flux characteristic. The positive effects of the simple design in the converter will result in less positive phenomena in the load. Also the voltage peaks are relatively large in comparison with the rated motor voltage. Whereas these effects can be handled in an effective way with some improvements for the motor f.i. up to 690 V, special measures have to be foreseen in medium voltage systems. Medium voltage converters need special attentions to overall voltage surges, flux characteristics and redundancies in the converter. Forced commutated converters are of multi-level type in medium voltage applications. Two main types of design principles are capable to combine the advantages of a voltage source inverter with the requirements of larger power outputs in the medium voltage range. One example for a five-level converter is given in fig. 2.11.

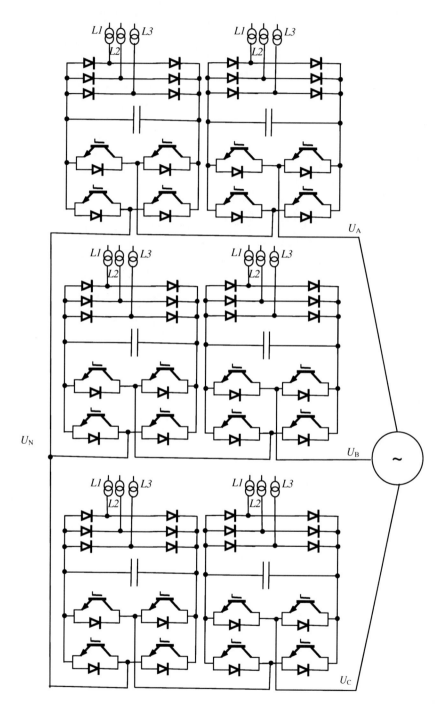

Fig. 2.11 Five-level cascaded converter topology acc. [2.9] – [2.12]

2.3 Forced Commutating Converters

The multi-level converter consists of several low voltage inverter bridges, which are connected over a diode rectifier to converter transformers. The transformer windings of each rectifier supply are separated galvanically. Each bridge can contribute towards the total voltage with one of three voltage levels. The d.c. link voltage is connected either positive or negatively. The third possibility is a direct short circuit of semi-conductive IGBTs and diodes at the load side. Two inverter bridges are connected in series in fig. 2.3. Five voltage levels can be realized between two times the negative and positive d.c. voltage of one bridge. Fig. 2.12 shows the measured phase voltage at motor terminals and its flux characteristic.

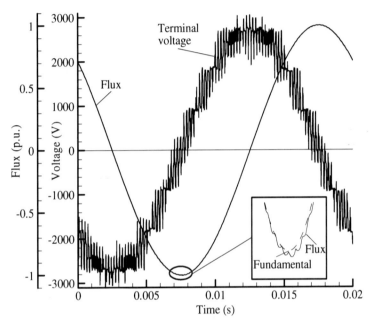

Fig. 2.12 Measured terminal voltage and flux characteristic of a medium voltage motor, which is fed by a multi-level cascaded converter

The flux characteristic is nearly sinusoidal. Voltage peaks are relatively small in comparison with the motors nominal voltage. Beside an improved voltage characteristic due to the multiple voltage levels, a higher combined pulse frequency can be realized, with the same individual frequency at each semiconductor. If both bridges are switched with a pulse frequency f.i. of 2 kHz an overall frequency of 4 kHz can be realized by a proper phase shift in the forced commutation. An alternative design of a multi-level converter is shown in fig. 2.13.

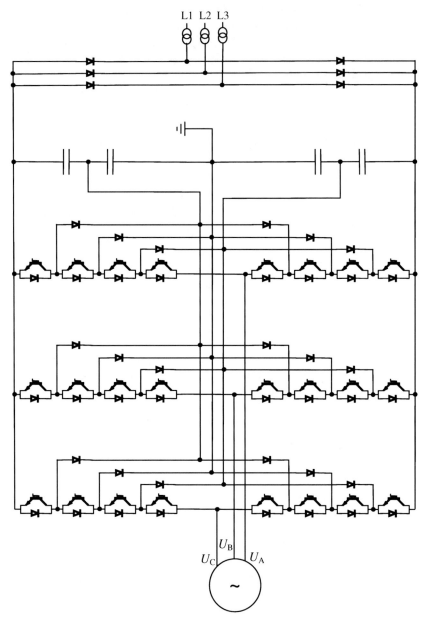

Fig. 2.13 Five-level diode clamped converter design acc. [2.9] – [2.12]

Multi-level diode clamped converters do grow in complexity with an increasing number of voltage levels. They are not as modular as the cascaded multi-level converter. Each level imposes a further capacitor in the d.c. link and a clamping

2.3 Forced Commutating Converters

by a diode. Due to directly series connected semi-conductive devices a strong control performance can be reached. They are often used in dynamic processes like in steel mills. A measured voltage characteristic for a three-level clamped diode converter is given in fig. 2.14.

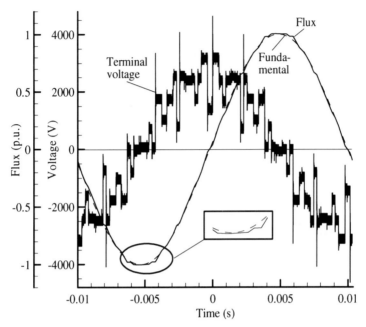

Fig. 2.14 Measured terminal voltage and flux characteristic of a medium voltage motor, which is fed by a three-level diode clamped converter

Voltage peaks have to be considered in comparison with the cascaded converter type. Nevertheless the flux characteristic is more sinusoidal than for the two-level converter.

Beside the bridge type, direct type converters do exist. These converters do connect the grid supply directly with the motor terminals without a d.c. link. One advantage of this converter type may be given by a reduction in converter losses, if IGBTs with reverse blocking capacity are used. Whereas the current flow in a bridge type of converter has to cross four semi-conductive elements, the current would only cross two elements in such a matrix converter. If normal standardized IGBTs are used one diode is connected in series and this advantage is gone. An example of a matrix converter with series connected diodes is given in fig. 2.15. The matrix converter type has two main disadvantages as well. The number of semi-conductive elements is given by twice the multiplication of the phases from the supply with the number of phases of the load. The supply is a three phase system and the load has three phases as well. Therefore eighteen semi-conductive elements are necessary.

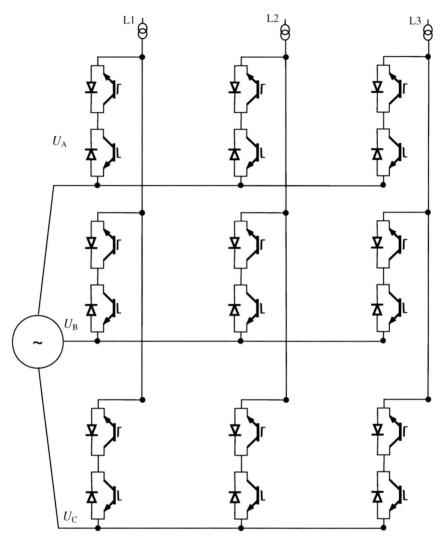

Fig. 2.15 Matrix converter for a three phase supply and three phase motor

The number of semi-conductive elements dominates the costs. The converter control is more complex than the straight forward pulse-width modulation in a converter with d.c. link.

2.4 Harmonic Content in the Voltage and Current Characteristics of Different Converter Types

The harmonic content in the voltage and current characteristics of the different converters will depend on the chosen control strategy. An overview and a detailed

2.4 Harmonic Content in the Voltage and Current Characteristics

description of the control strategies can be found in [2.13]. A classification for voltage source converters is given in fig. 2.16 according [2.13].

The feed forward methods pre-calculated pulse cycles and space vector-modulation are the most often applied strategies for larger pump and compressor applications. The power flow is always from the grid to the load. Fourier analysis of the voltage characteristic for these two control strategies are given in fig. 2.17. The analysis shows the harmonic content just for one specific operation point. The measurement for the pre-calculated pulse cycles has been taken at 17 Hz fundamental frequency, the space vector modulation at 66 Hz. Pre-calculated pulse cycles are stored in tables. A converter will get his pulse cycles out of different tables for different voltage or frequency ranges. The space vector modulation depends on the operation point as well. Pulse frequencies may change with the fundamental speed or are kept constant.

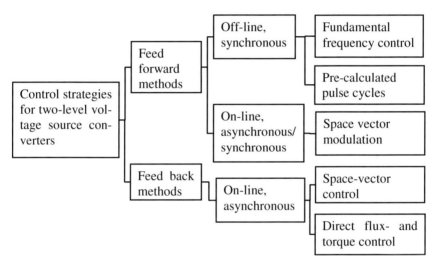

Fig. 2.16 Overview of different control strategies for two-level voltage source converters for drives, acc. [2.13]

The converter voltage characteristic shows harmonics of the third order. These are of common mode and will not create motor currents. Other harmonics up to the order nineteen have amplitudes of less than three percent in the pre-calculated controlled converter and less than one percent in the one with space vector modulation. The three-level converter control with pre-calculated pulse musters has more different harmonics than the characteristic of the two-level converter, which is space vector modulated, but the harmonics are smaller in amplitude. Mayor harmonics are the 41th, 49th and 59th. The 21th has an amplitude of about five percent of the fundamental. The harmonic content will change with other frequencies or voltages. The voltage characteristic of the two-level converter with space vector modulation is dominated by harmonics in line with the pulse frequency and it's multiple. First mayor harmonics are the 34th, 36th, 38th and 40th.

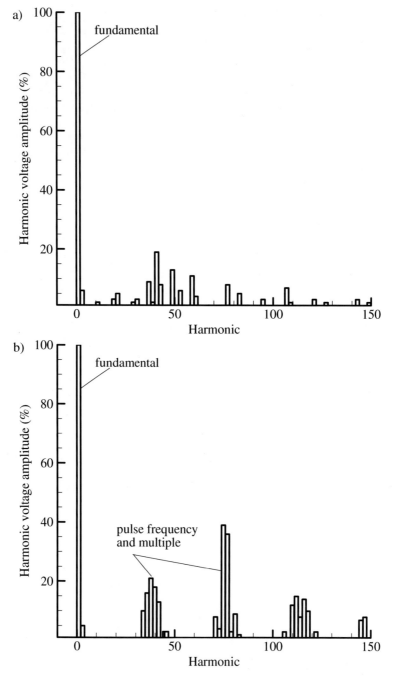

Fig. 2.17 a) Measured voltage harmonic content of a three-level converter with pre-calculated pulse cycles for square of current harmonic minimum and b) harmonic content for a two-level converter with space vector modulation

2.4 Harmonic Content in the Voltage and Current Characteristics

An extreme low amount of harmonics will occur in the characteristic of the multi-level medium voltage converter, see fig. 2.18. The first harmonics have on order of ninety-five or higher. Even these harmonics, which are caused according to the pulse frequency, are below five percent. The analysis is based on a measurement at 50 Hz.

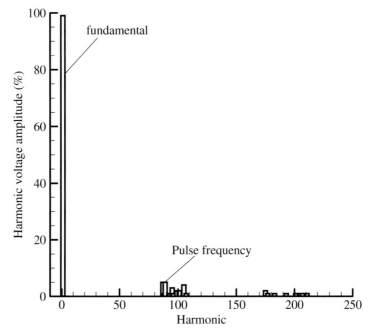

Fig. 2.18 Measured voltage harmonic content within a multi-level medium voltage converter with low voltage cells

Chapter 3
Calculation-Methods for Converter Fed Electrical Machines

3.1 Analytical Calculation Methods for Special Operation Points and Machine Model Integration into Circuits

Different basis quantities of asynchronous machines need to be calculated within a drive application. The machine current, stalling torque, losses and temperatures do belong to these quantities. Converters are in principle capable to create a relatively sinusoidal current if they use an appropriate converter design and pulse frequency, but in order to save costs the supplied voltage form may be far beyond any sinusoidal supply. In both cases it is often sufficient to determine the losses and temperature increase based upon the current of the first harmonic and add an appropriate temperature margin for the higher harmonics. The machine currents are derived from two voltage equations 3.1, 3.2, acc. [3.1]:

$$\underline{U}_1 = R_1 \underline{I}_1 + j\omega \left[L_1 (1 + \sigma_{d1}) + L_{N1} + L_{S1} \right] \underline{I}_1 + j\omega 2 p M_{2,1} \underline{I}_{R,2}, \tag{3.1}$$

$$\underline{U}_2 = \left[2R_{R2} + R_{S2} \left(2\sin\left(p\frac{2}{Z_2} \cdot \frac{\pi}{2} \right) \right)^2 \right] \underline{I}_{R,2}$$

$$+ js\omega \left[L_2 (1 + \sigma_{d2}) + 2L_{R2} + L_{N2} \left(2\sin\left(p\frac{2}{Z_2} \cdot \frac{\pi}{2} \right) \right)^2 \right] \underline{I}_{R2} + js\omega M_{1,2} \underline{I}_1 \tag{3.2}$$

The stator voltage $\underline{U}_1$ and current $\underline{I}_1$ are phase quantities. The stator consists of m phases. The number of phases within the rotor is given by the rotor slots Z_2. One loop is assumed to be created by two adjacent bars. The loop-current is given by the ring-current $\underline{I}_{R,2}$. The bar-current $\underline{I}_{St,2}$ is given by the difference between two adjacent ring-currents:

$$\underline{I}_{St,2} = 2 \cdot \underline{I}_{R,2} \cdot \sin\left(p\frac{\pi}{Z_2} \right) \cdot \left(\pm j \cdot \exp\left(\pm j\frac{\pi \cdot p}{m} \right) \right). \tag{3.3}$$

The ± sign describe the phase difference between the two bars, which are adjacent to the part of the ring. The calculation of the individual inductivities is presented in attachment A.1 for the linear case. Stator and rotor main inductivities are given by L_1 and L_2. Mutual inductivities are described by $M_{1,2}$ and $M_{2,1}$. These inductivities are not those of single circuits, but include the influence of the other phases in the rotor or stator. Three leakage components are considered with the slot leakage L_{N1}, L_{N2}, the end winding or ring leakage L_{S1}, L_{R2} and the air-gap leakage, which is given by the factor σ_{d1}, σ_{d2}. The air-gap leakage represents the leakage influence of higher harmonics.

Analytical calculation methods, which consider the non-linearity of iron and the calculation of iron losses, do exist for instance in [3.2]. These calculation methods are quite similar to numerical field calculations. They are iterative as well and can be considered as an alternative method to determine the inductances in equations 3.1 and 3.2.

Main loss components within asynchronous machines are given by rotor and stator conduction losses acc. equation 3.4 and 3.5.

$$P_{cu,1} = m \cdot |\underline{I}_1|^2 \cdot R_1 \tag{3.4}$$

$$P_{cu,2} = Z_2 \cdot |\underline{I}_{R,2}|^2 \cdot \left(2 \cdot R_{R2} + R_{S2} 4 \sin^2 \left(p \frac{\pi}{Z_2} \right) \right) \tag{3.5}$$

Beside the losses within the windings itself, iron, friction and additional losses contribute strongly towards the machines' loss balance.

The asynchronous torque of the base frequency is direct proportional to the rotor losses:

$$M_{asyn} = \frac{p}{s \cdot \omega_{el}} \cdot P_{cu,2}, \tag{3.6}$$

with the slip:

$$s = \frac{\omega_{el} / p - \omega_{mec}}{\omega_{el} / p}. $$

The stalling torque is given acc. [3.3] to:

$$M_S \approx \frac{m}{\omega_{el}} \frac{U_1^2}{2 \left(R_1 \pm \sqrt{R_1^2 + \omega_{el} L_{1\sigma}^2 + \omega_{el} L_{2\sigma}^{'2}} \right)}. \tag{3.7}$$

The influence of the resistance is relatively small in large asynchronous machine for 50Hz direct on line operations. In case of larger frequencies of 200Hz the term may gets more important due to eddy current phenomena. The leakage inductance in the rotor is given in equation 3.8. The one for the stator can be derived from equation 3.1 analog:

$$L_{2\sigma}' = \frac{m}{Z_2} \cdot \left(\frac{w_1}{w_2} \cdot \frac{k_{w1}}{k_{w2}} \right)^2 L_{2\sigma}$$

$$= \frac{m}{Z_2} \cdot \left(\frac{w_1}{w_2} \cdot \frac{k_{w1}}{k_{w2}} \right)^2 \cdot \left(L_2(\sigma_{d2}) + 2L_{R2} + L_{N2} \left(2 \sin \left(p \frac{2}{Z_2} \cdot \frac{\pi}{2} \right) \right)^2 \right) \tag{3.8}$$

3.1 Special Operation Points and Machine Model Integration into Circuits

The stalling torque is often in the field weakening area the limiting factor for the asynchronous machine. The stator current, which can be calculated out of equation 3.1 and 3.2 is needed for a proper converter dimensioning. In compressor or pump applications the motor temperature becomes the main design criteria. The temperature is mainly determined by the copper losses, if a known machine design is used and iron losses are kept constant.

Depending on the kind of converter and motor in some cases even more than 20K reserve has to be taken into account due to those losses, which are imposed by the converter non-sinusoidal supply. The machine voltage can be considered as the summation of its harmonics for the harmonic current calculation. An example of the harmonic content within the terminal voltage of a low voltage 690V converter is given in fig. 3.1.

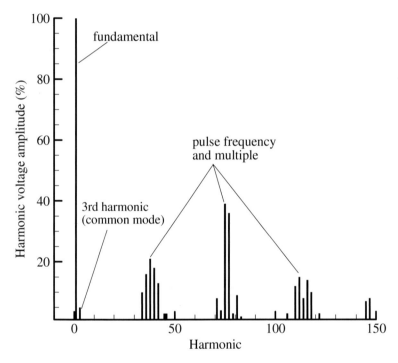

Fig. 3.1 Fourier-analysis of the terminal voltage from a low voltage variable frequency drive with vector modulation

The voltage harmonics can be applied to equations 3.1 and 3.2 in order to get the harmonic current content. Whereas the operational flux point will dominate the amount of iron saturation and therefore the main inductivities, the eddy current losses in the stator winding depend strongly upon the current frequency and have to be considered in the applied stator resistance. Additionally the rotor parameters have to be recalculated for the slip of each harmonic.

A similar approach can be applied for synchronous machines. The equation system of equation 3.1 and 3.2 has to be extended by the excitation winding of the

30   3 Calculation-Methods for Converter Fed Electrical Machines

rotor, if a magnetic symmetrical rotor is assumed. If asymmetries between the neutral zone and the pole zone exist or the coupling between damper and exciter winding must be taken into account, six equations according [3.4] are necessary. The magnetic flux distribution within a synchronous machine does rotate synchronous with the rotor for the fundamental supply frequency. Asynchronous operation has to be considered for higher voltage harmonics, which induce currents in the damper cage.

The equation system, which considers all three circuits is given by equations 3.9-3.12:

$$\underline{U}_1 = R_1 \underline{I}_1 + j\omega \left[L_1 (1 + \sigma_{d1}) + L_{N1} + L_{S1} \right] \underline{I}_1 + j\omega 2 p M_{2,1} \underline{I}_{R,2} + j\omega_{el} {}^{\nu}\underline{M}_{1,f} I_f , \quad (3.9)$$

$$0 = \left[2R_{R2} + R_{S2} \left(2\sin\left(p \frac{2}{Z_2} \cdot \frac{\pi}{2} \right) \right)^2 \right] \underline{I}_{R,2}$$

$$+ js\omega \left[L_2 (1 + \sigma_{d2}) + 2L_{R2} + L_{N2} \left(2\sin\left(p \frac{2}{Z_2} \cdot \frac{\pi}{2} \right) \right)^2 \right] \underline{I}_{R2} + js\omega M_{1,2} \underline{I}_1 \qquad , (3.10)$$

$$U_f = R_f I_f , \qquad (3.11)$$

with

$$ {}^{\nu}\underline{M}_{1,f} = {}^{p}M_{1,f} \cdot \exp j \left(\vartheta_0 - \frac{\pi}{2} \right), \nu = p \qquad (3.12)$$

$$ {}^{\nu}\underline{M}_{1,f} = 0, \nu \neq p$$

The calculation of the mutual inductivity $M_{1,f}$ between the stator and the field winding as well as of the field resistance R_f is presented in attachment A.2. The mutual inductivity as well as stator voltage and current change sinusoidal in time for constant speed. Only the complex amplitudes are calculated. Equation 3.9 considers the induced voltage by currents of the damper cage and by the field winding. In equation 3.10 and 3.11 the coupling between the field winding and the damper cage is neglected. Indeed the converter harmonics impose in the rotor cage currents of frequencies with more than 100 Hz causing a small skin depth. Hardly any voltage is induced in the field winding with this frequency. Therefore the equations are valid for the stator voltage fundamental and each converter voltage harmonic in principle. The induced voltage due to the exciter field does not contribute to higher stator voltage harmonics nor do damper currents of higher harmonics induce voltages for the fundamental. No damper currents are induced for the fundamental, because the slip is zero.

Higher harmonic stator and rotor currents can be calculated based on equations 3.9 to 3.12. The equations can also be integrated in mathematical circuit models, which are coupled to semi-conductive elements. Active and reactive power of the synchronous machine is given by equations 3.13 and 3.14:

3.1 Special Operation Points and Machine Model Integration into Circuits

$$P_{\mathrm{w}} = 3\frac{U_1 E}{X_{\mathrm{d}}}\sin\vartheta_0 , \tag{3.13}$$

$$P_{\mathrm{b}} = 3\left(\frac{U_1^2 - EU_1\cos(\vartheta_0)}{X_{\mathrm{d}}}\right), \tag{3.14}$$

with

$$E = \omega_{\mathrm{el}}{}^{\mathrm{p}}M_{1,\mathrm{f}}I_{\mathrm{f}} . \tag{3.15}$$

The active power defines the mean power of the fundamental harmonics and the reactive power the power oscillation of the fundamental. If the required active- and reactive power are known, the exciter current can be derived out of the equations 3.13-3.15.

Beside scientifically exact machine equations, the converter current may be determined by a more basic approach. The load of the driven machine may be given by P_{load}. The converter current I_{conv} can be approximated by equation 3.16:

$$I_{\mathrm{conv}} = \frac{P_{\mathrm{load}}}{\cos\varphi \cdot \eta \cdot \sqrt{3} \cdot U_{\mathrm{uv}}} \tag{3.16}$$

Typical values for the power factor and efficiency for asynchronous machines with copper cage and synchronous machines are given in table 3.1 based on the experience of several hundred machines:

Table 3.1 Power factor and efficiencies for electrical machines in the range of 1MW-50MW

Machine type	Asynchronous no. of poles: 10...2	Synchronous	Synchronous with permanent magnets
Power factor $\cos\varphi$	0.84...0.90	0.95...1.00	0.7...0.9
Efficiency η	93%...97.3%	97.5%...98.5%	98.0%-99%

Additional losses due to higher harmonics can be taken into account by a reduction of the machine performance up to 20% for converters with highly utilized, low pulse frequency power electronic elements and basic converter architecture. Multi-level converters allow for a very similar performance than for sinusoidal supply.

A basic choice of the motor converter system is generally possible with the basic approach of equation 3.16. A deeper understanding of the individual phenomenon will not occur. Especially the consideration of capacitive effects, the plant design or the commutation of semi-conductive devices requires more sophisticated models. These models are mainly based on the voltage equations of the electrical machines. Equation 3.1 and 3.2 for the asynchronous machine as well as the equation 3.9 – 3.12 for the synchronous machine can be integrated in a mathematical circuit model. The equations can be applied either directly or by circuit diagrams, which couple the rotor directly to the stator. The rotor equations have to be transformed towards the stator side similar to the transformer theory [3.3] in this case.

Circuit models of electrical systems are always based on a lump- and admittance-matrix [3.5]. The circuit model is split in several single branches, which describe the individual elements like diodes, thyristors or even complete motors. Each branch consists of a certain relationship between voltage and current. This relationship may be non-linear or even controlled by an additional control system in case of power electronics [3.6, 3.7]. A circuit model is a most appropriate method to determine terminal quantities and to simulate converter performance. In order to have a deeper look within the electrical machine other methods, which will be described in detail in the following chapters, numerical methods are superior. Therefore only a short overview is given. The mathematical models can be derived from literature [3.5, 3.8].

One of the main challenges within the development of electrical drives is an optimized control strategy of the applied power electronics. A relative simple circuit model of the electrical machine is sufficient in order to dimension the main converter components in line with the chosen control strategy. One of the control targets is to provide an acceptable sinusoidal first harmonic with a relative low pulse frequency to reduce converter losses. Generally low pulse frequencies increase additional losses and noise within electrical machines. The integration of the mathematical machine description in a circuit model with a transient solver allows for transient solutions. Even though phenomena as additional losses nor noise can not be calculated without a more sophisticated machine approach, terminal quantities like stator currents or voltages can be derived with sufficient accuracy.

Each circuit topology will be given by branches and nodes. Along any branch a voltage will be determined by a function, which depends upon the current within the branch or in more complex situations like in electrical machines it will depend on currents within other branches as well. The mathematical description of the electrical circuits is based on the following equation system [3.5, 3.8]:

- Branch characteristic:

$$
\begin{pmatrix} U_1 \\ \cdot \\ \cdot \\ \cdot \\ U_i \\ \cdot \\ \cdot \\ \cdot \\ U_n \end{pmatrix} = \begin{bmatrix} Z_{11} & \cdot & \cdot & \cdot & Z_{1i} & \cdot & \cdot & \cdot & Z_{1n} \\ \cdot & \cdot & \cdot & \cdot & \cdot & \cdot & \cdot & \cdot & \cdot \\ \cdot & \cdot & \cdot & \cdot & \cdot & \cdot & \cdot & \cdot & \cdot \\ \cdot & \cdot & \cdot & \cdot & \cdot & \cdot & \cdot & \cdot & \cdot \\ Z_{i1} & \cdot & \cdot & \cdot & Z_{ii} & \cdot & \cdot & \cdot & Z_{in} \\ \cdot & \cdot & \cdot & \cdot & \cdot & \cdot & \cdot & \cdot & \cdot \\ \cdot & \cdot & \cdot & \cdot & \cdot & \cdot & \cdot & \cdot & \cdot \\ \cdot & \cdot & \cdot & \cdot & \cdot & \cdot & \cdot & \cdot & \cdot \\ Z_{n1} & \cdot & \cdot & \cdot & Z_{ni} & \cdot & \cdot & \cdot & Z_{nn} \end{bmatrix} \cdot \begin{pmatrix} I_1 \\ \cdot \\ \cdot \\ \cdot \\ I_i \\ \cdot \\ \cdot \\ \cdot \\ I_n \end{pmatrix}, \qquad (3.17)
$$

3.1 Special Operation Points and Machine Model Integration into Circuits 33

- First Kirchhoff-Law for m-1 independent nods:

$$\begin{pmatrix} 0 \\ 0 \\ 0 \\ 0 \\ 0 \\ 0 \\ 0 \\ 0 \end{pmatrix} = \begin{bmatrix} 1 & \cdots & -1 & \cdots & 0 & 1 \\ & \cdots & \cdots & \cdots & \cdots & \cdots \\ & \cdots & \cdots & \cdots & \cdots & \cdots \\ & \cdots & \cdots & \cdots & \cdots & \cdots \\ 0 & \cdots & 1 & \cdots & -1 & 0 \\ & \cdots & \cdots & \cdots & \cdots & \cdots \\ & \cdots & \cdots & \cdots & \cdots & \cdots \\ & \cdots & \cdots & \cdots & \cdots & \cdots \\ & \cdots & 1 & \cdots & 1 & -1 \end{bmatrix} \cdot \begin{pmatrix} I_1 \\ \cdot \\ \cdot \\ \cdot \\ I_i \\ \cdot \\ \cdot \\ I_{m-1} \\ I_m \end{pmatrix} , \qquad (3.18)$$

- Second Kirchhoff-Law for n-m-1 independent loops:

$$\begin{pmatrix} 0 \\ \cdot \\ \cdot \\ \cdot \\ 0 \\ \cdot \\ \cdot \\ \cdot \\ 0 \end{pmatrix} = \begin{bmatrix} 1 & \cdots & 0 & \cdots & 0 \\ & \cdots & \cdots & \cdots & \cdot \\ & \cdots & \cdots & \cdots & \cdot \\ & \cdots & \cdots & \cdots & \cdot \\ 1 & \cdots & 1 & \cdots & -1 \\ & \cdots & \cdots & \cdots & \cdot \\ & \cdots & \cdots & \cdots & \cdot \\ & \cdots & \cdots & \cdots & \cdot \\ 0 & \cdots & -1 & \cdots & 1 \end{bmatrix} \cdot \begin{pmatrix} U_1 \\ \cdot \\ \cdot \\ \cdot \\ U_i \\ \cdot \\ \cdot \\ \cdot \\ U_{n-m-1} \end{pmatrix} . \qquad (3.19)$$

Equations 3.17 to 3.19 will define an equation system with 2n-1 equations. One voltage potential can be chosen arbitrary. Any control algorithm can be implemented in the branch characteristic Z_{ij}. Even though the mathematical model of an electrical circuit is principally defined by these equation systems, special methods are needed for an adequate choice of loops [3.5, 3.8].

A typical example for this circuit approach is given in fig. 3.2 for a brushless excitation system with rotating diode rectifier.

The exciter is modelled as a synchronous machine, which is based on the equation system 3.9-3.12. The machine is built as a machine, where the three phase armature is rotating, whereas the three phase exciter winding is fed by a d.c. current. The rotating armature winding is connected to a diode bridge, which rectifies the a.c. voltage to a d.c. voltage with some ripple of higher frequency. The

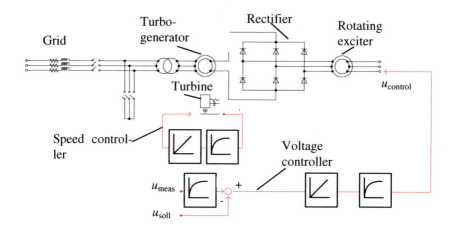

Fig. 3.2 Circuit diagram for a rotating brushless excitation system with diode rectifier

saturation of the main flux part within the excitation machine is taken into account. This is important especially due to the limited ceiling capacity of the excitation machine. A simulation model like in fig. 3.2 has to be verified by measurements. Comparisons with measurements are given in fig.3.3. The measurements have been done on a generator- gas-turbine arrangement with a rated output of 69 MVA. The generator has been started by a static frequency converter together with the gas-turbine. Afterwards it has been excited to provide 29 MVAr of reactive power. A rejection of this power causes an increase in the terminal voltage, fig. 3.3 a). The automatic voltage regulator has the task to control the terminal voltage to nominal voltage again [3.10]-[3.12]. Therefore the field voltage of the brushless exciter is decreased directly after the rejection to zero. Afterwards it is increased to the value of no load excitation. The exciter field current follows this voltage change, fig. 3.3 b).

Measurement and calculation agree well for the terminal voltage. The measured brushless exciter field current deviates stronger. On the one hand the measurement has been filtered to the mean current value, on the other hand the measured excitation current rises slightly over the no load current, whereas the calculation is more aperiodic.

If the circuit modulation of the electrical machine is replaced by numerical field calculation and coupled to semi-conductive circuit elements the principle handling of the converter circuit topology will be the same. A detailed description of the equation system on an example circuit will be given in the following chapter.

3.2 Non-linear Transient Time-Stepping Numerical Field Calculation

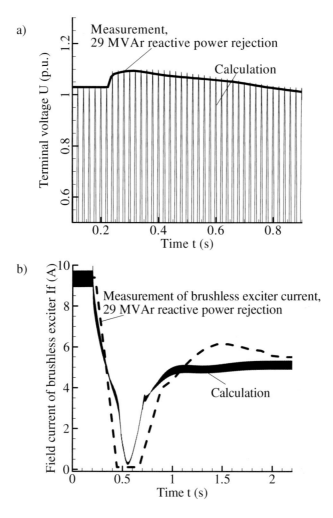

Fig. 3.3 Comparison of the measured and calculated generator terminal voltage a), and brushless exciter stator current b), calculation has been done with [3.9]

3.2 Non-linear Transient Time-Stepping Numerical Field Calculation with Integrated Circuit Elements

Several numerical methods can be applied for the calculation of electrical machines [3.13]-[3.23]. The finite element method as well as finite difference method can be applied on the same kind of electromagnetic problem in principle. One of the simplest cases of field calculation is the linear case with impressed d.c.-currents. Even though some details or inductivities in electrical machines can be determined by this kind of mathematical model, much more elaborated models are necessary to get correct results especially for machines, which are connected to power electronics. Here the winding is connected to a rectifier or switch, which causes transients in the winding during switching. A non-linear numerical time-stepping method with rotating rotor and power electronic circuit model has to be applied [3.14]-[3.16], [3.18, 3.19, 3.28]. Basic theories for the combination between the circuit theory and field theory are summarized in one mathematical model. This is possible by the application of finite element or finite difference time-stepping methods. Both methods are quite similar in general concerning their practical application. The finite element methods have the advantage to be easier adjustable to complex geometries. On the other hand the finite difference method is easier to implement and has shown good results in comparison with measurements due to its most often inherently applied regular grid in the air-gap. The saturation effect is also well understood and has often been simulated in the past [3.14, 3.15]. A further step in numerical field solutions is its simultaneous combination with circuit models. Its' possible application to simple winding structures within the finite element model has been shown in [3.15, 3.16]. These methods are elaborated further, so that it is possible to apply the numerical solution to complex circuit models like multi-phase polygonal connected brushless exciters including the rectifier model as a network of many electrically non-linear branch elements [3.24] – [3.26]. The non-linear transient time-stepping numerical field calculation method with integrated circuit elements is in the following shown in detail on the example of finite differences. It is possible to transfer any two-dimensional motor geometry to a mathematical model applying numerical means. The geometric discretization is realized in line with the finite difference method and the geometry is covered by an orthogonal network of several thousands of nodes. The discretization in time is based upon the Crank-Nicholson procedure or the modified trapezoidal rule, Θ-method [3.15], respectively.

The magnetic vector potential $\vec{A}$ with its definition $\vec{B} = rot\,\vec{A}$ is used for the calculation of electromagnetic fields in general. In two-dimensional cases only the axial vector-potential component exists. Based on the introduction of the vector-potential the Maxwell equations are summarized within the eddy current equation considering external fields as well. The eddy current equation for two-dimensional problems can be generally written as:

$$\Delta A_z = +\mu\chi\left(\frac{\partial A_z}{\partial t} + e_z \cdot \text{grad}\,\varphi\right)\,, \qquad (3.20)$$

where χ stands for the electrical conductivity and μ for the absolute magnetic permeability. The unit vector in axial direction is denoted $\mathbf{e}_z$.

The merely axial electric potential gradient grad φ could be expressed by a source current density $J_S = -\,\text{grad}\;\varphi$. But this might be misleading [3.17] as such a source could be mixed up with external current sources, which e.g. would be zero for an open loop massive conductor circuit without grad φ being zero. Grad φ vanishes only for regions with an ideal short-circuit return path, but has to be taken into account as an additional unknown for solid conductors carrying an impressed

3.2 Non-linear Transient Time-Stepping Numerical Field Calculation

current or being part of an external circuit, e.g. the end ring in a squirrel cage induction machine.

The vector potential A_z is calculated in each node of the finite difference grid, fig. 3.4.

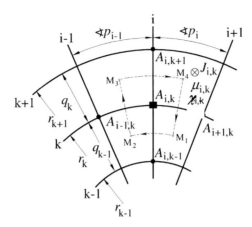

Fig. 3.4 Detail of a finite difference mesh with the path of integration via the cell median points M1 ... M4 around the central node (i, k)

The current density $J_{i,k}$, the magnetic permeability $\mu_{i,k}$ and the electrical conductivity $\chi_{i,k}$ are indicated for the adjacent grid cell (i, k) as domain properties. The differential equation 3.20 is converted into a system of algebraic equations 3.21 by evaluating Ampere's law for each node along the path of integration in fig. 3.4 and replacing differential quotients by difference quotients of the adjacent nodal vector potentials and distances. The use of the Crank-Nicholson scheme finally yields for ideally short-circuited eddy current regions:

$$\frac{h}{2}(MA_{i,k} - NA_{i,k+1} - SA_{i,k-1} - OA_{i+1,k} - WA_{i-1,k})\big|_{t+h} + F_{\chi,i,k} A_{i,k}\big|_{t+h}$$
$$= -\frac{h}{2}(MA_{i,k} - NA_{i,k+1} - SA_{i,k-1} - OA_{i+1,k} - WA_{i-1,k})\big|_{t} + F_{\chi,i,k} A_{i,k}\big|_{t} \quad (3.21)$$

The reluctance coefficients N, O, S and W and the conductance coefficients $F_{\chi,i,k}$ in 3.21 are calculated depending on the material properties and the mesh geometry according to:

$$F_{\chi,i,k} = \left(\frac{q_k}{2}\left(r_k + \frac{q_k}{4}\right)(p_i \chi_{i,k} + p_{i-1} \chi_{i-1,k})\right) + \\ + \left(\frac{q_{k-1}}{2}\left(r_k - \frac{q_{k-1}}{4}\right)(p_i \chi_{i,k-1} + p_{i-1} \chi_{i-1,k-1})\right), \quad (3.22)$$

$$M = N + S + W + O,$$

$$N = \frac{1}{4}\frac{r_{k+1} + r_k}{q_k}\left(\frac{p_i}{\mu_{i,k}} + \frac{p_{i-1}}{\mu_{i-1,k}}\right), \tag{3.23}$$

$$S = \frac{1}{4}\frac{r_{k-1} + r_k}{q_{k-1}}\left(\frac{p_i}{\mu_{i,k-1}} + \frac{p_{i-1}}{\mu_{i-1,k-1}}\right),$$

$$O = \frac{1}{p_i}\left(\mu_{i,k-1}\ln\left(\frac{2r_k}{r_{k-1} + r_k}\right) + \mu_{i,k}\ln\left(\frac{r_{k+1} + r_k}{2r_k}\right)\right), \tag{3.24}$$

$$W = \frac{1}{p_{i-1}}\left(\mu_{i-1,k-1}\ln\left(\frac{2r_k}{r_{k-1} + r_k}\right) + \mu_{i-1,k}\ln\left(\frac{r_{k+1} + r_k}{2r_k}\right)\right), \tag{3.25}$$

with r_k: radius of arc no. k in meters (radial discretisation),

$$q_k = r_k - r_{k-1},$$

φ_i: angular coordinate of ray no. i in radians (circumferential discretisation),

$$p_i = \varphi_i - \varphi_{i-1}.$$

The implementation of the method, which considers eddy currents, is not foreseen for the rotating armature, where the winding is connected to external circuitry and treated as a 'filamentary problem' [3.18]. The coupling between the rotating FD network with its stationary counterpart is carried out by a special FD-star formulation [3.14], which is essentially comparable to a mesh regeneration, the simplest coupling method in FEM [3.18], also referred to as moving band method (MBM) [3.19].

Fig. 3.5 shows an example of an electrical machine which is connected to a circuit consisting mainly of power electronic elements. Indeed the power electronic circuit builds up the diode rectifier of a brushless exciter. The complete circuit of the $m = 9$ polygonal-connected machine windings, the rotating 18-pulse bridge rectifier and the generator field winding is given.

The grey marked machine windings are modeled in the finite difference Time-stepping scheme according to their ohmic-inductive voltage equations implying Faraday's law of induction. The equation for voltage $u_{w,i}$ and current $i_{w,i}$ of winding no. i can be written as:

$$u_{w,i} = R_w \cdot i_{w,i} + \frac{d}{dt}\Psi_{w,i} = R_w \cdot i_{w,i} + \frac{d}{dt}\left[L_{\sigma,w}i_{w,i} + \right.$$

$$\left. + N_{turn,w}\, l_{Fe} \cdot \left(\frac{1}{a_{CS+,w}}\iint_{CS+,w}A\,da - \frac{1}{a_{CS-,w}}\iint_{CS-,w}A\,da\right)\right] \tag{3.26}$$

3.2 Non-linear Transient Time-Stepping Numerical Field Calculation

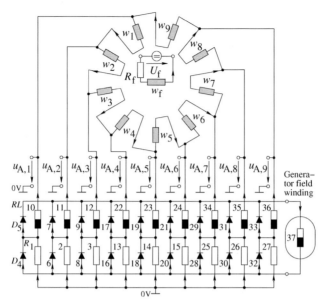

Fig. 3.5 Circuit scheme of the machine windings $w_{1...9}$ and w_f, the rectifier circuits and the d.c. load of the turbo-generator field winding

The flux linkage $\psi_{w,i}$ is given by the difference of average vector potential values on the positive and negative winding cross-sections $a_{cs+,w}$ and $a_{cs-,w}$ multiplied by the iron length l_{Fe} and the number of winding turns $N_{turn,w}$. A leakage flux represented by the inductance $L_{\sigma,w}$ for the winding end zone can be taken into account additionally.

The flux linkage integral is transformed into a sum according to the spatial discretization of the machine cross-section. The nodal vector potentials on the winding cross-sections are weighted by factors $F_{w,j}$ $F_{w,j}$ according to the turn density N_{turn}/a_{cs} and the nodal adjacent area. Discretization in time with a step h is carried out by using the θ-method, equation 3.27.

$$-\sum_{j(CS,w+)} F_{w,j} A_j \Big|_{t+h} + \sum_{j(CS,w-)} F_{w,j} A_j \Big|_{t+h}$$
$$+\frac{1-\theta}{l_{Fe}} h \cdot u_{w,i} \Big|_{t+h} - \frac{L_{\sigma,w} + (1-\theta) h R_w}{l_{Fe}} i_{w,i} \Big|_{t+h} =$$
$$= -\sum_{j(CS,w+)} F_{w,j} A_j \Big|_t + \sum_{j(CS,w-)} F_{w,j} A_j \Big|_t - \frac{\theta h}{l_{Fe}} u_{w,i} \Big|_t - \frac{L_{\sigma,w} - \theta h R_w}{l_{Fe}} i_{w,i} \Big|_t$$

(3.27)

with

$$F_{w,j} = \frac{1}{4} \frac{N_{turn}}{a_{cs}} \left[\left(r_k + \frac{q_k}{4} \right) \cdot (p_i \cdot q_k + p_{i-1} \cdot q_k) + \left(r_k - \frac{q_{k-1}}{4} \right) \cdot (p_i \cdot q_{k-1} + p_{i-1} \cdot q_{k-1}) \right] \cdot$$

40   3 Calculation-Methods for Converter Fed Electrical Machines

This is equivalent to the trapezoidal rule of integration for $\theta = 0.5$. θ as defined here should not exceed 0.5 to be numerically stable, but can be chosen slightly below in order to provide damping for spurious numerical oscillations.

The machine model is linked to the external circuit which is made of lumped elements with $m = 9$ line-to-ground voltages $u_{A,1...9}$. These occur as common variables of the FD-model and the circuit equations. The following mesh equations are set up for a polygonal connection of the machine windings. They are directly embedded in 3.27, when the system of equations is created. Here the winding voltages $u_{w,i}$ are replaced:

$$u_{w,i}\big|_{t+h} = \begin{cases} u_{A,i}\big|_{t+h} - u_{A,i+1}\big|_{t+h}, & i = 1 \ldots m-1 \\ u_{A,m}\big|_{t+h} - u_{A,1}\big|_{t+h}, & i = m \end{cases}$$

(3.28)

The rectifier network consists of $N_b = 37$ branch elements. The resistors R_1, R_1, R_1, R_2, R_3, R_{13}, R_{14}, R_{15} and R_{25}, R_{26}, R_{27} are inserted in order to provide a symmetrical common grounding of the whole arrangement, which on the one hand is necessary as a potential reference point for the machine terminal voltages $u_{A,1...9}$ and on the other hand improves numerical stability. A relatively high but numerically still relevant value of $R = 10M\Omega$ is chosen for them in order to avoid any influence on the current distribution in the network. The algebraic equation to be solved for an instant of time $t + h$ is given by Ohm's law only.

The bridge rectifier itself consists of the diodes D_4, D_6, D_8, D_{16}, D_{18}, D_{20}, and D_{28}, D_{30}, D_{32} for negative and D_5, D_7, D_9, D_{17}, D_{19}, D_{21}, and D_{29}, D_{31}, D_{33} for positive voltage half-cycles. They are connected to the machine windings via ohmic-inductive elements RL_{10}, RL_{11}, RL_{12}, RL_{22}, RL_{23}, RL_{24} and RL_{34}, RL_{35}, RL_{36}. The generator field winding at the rectifier d.c. terminals is modeled as an ohmic-inductive branch RL_{37}. The time-discretized voltage-current relation for an element no. j with inductance L_j and resistance R_j is:

$$u_j\big|_{t+h} - \left(R_j + \frac{L_j}{(1-\theta)h}\right)i_j\big|_{t+h} =$$

$$= -\frac{\theta}{1-\theta}u_j\big|_t + \left(\frac{\theta}{1-\theta}R_j - \frac{L_j}{(1-\theta)h}\right)i_j\big|_t$$

(3.29)

The diodes are described as static non-linear branch elements according to an exponential voltage-current characteristic with the reverse current I_r, an emission-constant m_ε and the temperature voltage U_T:

$$i_{D,j}\big|_{t+h} - I_r \cdot \left(\exp\left(\frac{u_{D,j}\big|_{t+h}}{m_\varepsilon U_T}\right) - 1\right) = 0,$$

(3.30)

with $U_T = kT$, k being Boltzmann's constant and T the thermodynamic temperature in Kelvin.

3.2 Non-linear Transient Time-Stepping Numerical Field Calculation

The non-linearity requires iterative treatment during the solution process as does the magnetic saturation of the field model. The conducting state of 3.30 is linearized by the differential resistance of the diode and a voltage source with respect to the preliminary solution in each iterative step, i.e. by the tangent to the characteristic. In the non-conducting state an absolute conductance i_D/u_D is used instead. Each aforementioned branch element contributes one equation for the current voltage distribution according to 3.26, 3.29, 3.30 or the mere ohmic law. The implementation applied here successfully utilizes an ad hoc approach in contrast to sophisticated reductive methods like the loop current method with solid conductor and capacitor voltages [3.17] or hybrid methods based on tree branch voltages and co-tree branch currents [3.21]. All branch quantities are included in one overall system of equations without respect to retain symmetry or other special properties. This approach for network analysis of keeping more network quantities than principally necessary as state variables is also known as the "sparse tableau approach" as referenced in [3.27]. The number of network equations appears to be negligible in the presence of thousands of field equations. Moreover, ill-conditioned systems, which would result from practically cutting or short-circuiting branches by extreme values of their resistances in reductive methods, can be avoided by the sparse tableau approach.

Structural similarities of the resulting system matrix with those presented in [3.21] and [3.22] can nevertheless be observed concerning the field equation block and the coupling terms. The lack of structure is restricted to the network block only. Robustness of the block-wise direct Gaussian solution process is ensured by a total pivot search, when eliminating the unstructured network block.

The contribution to the network block given by the topology of the whole network including the machine windings of the field model is described by Kirchhoff's well-known laws listed below in detail for this exciter arrangement.

According to fig. 3.5 the rectifier network obviously contains 11 internal nodes, 9 diode bridge nodes and the two d.c. terminals. Moreover there are 9 machine terminals. The nodal current equation 3.31 of the bridge nodes can be denoted according to the branch numbering in fig. 3.5:

$$\left(-i_{12 \cdot l+k} - i_{12 \cdot l+2 \cdot (k+1)} + i_{12 \cdot l+2 \cdot (k+1)+1} + i_{12 \cdot l+k+9} \right)\Big|_{t+h} = 0, \tag{3.31}$$

for $k = 1,2,3$ and $l = 0,1,2$.

The following holds for the d.c. terminals:

$$\left(-\sum_{l=0}^{2}\sum_{k=1}^{3} i_{12 \cdot l+2 \cdot (k+1)+d} + i_{37} \right)\Bigg|_{t+h} = 0, \tag{3.32}$$

for $d = 0,1$.

The polygonal connection of the windings has to be taken into account for the machine terminals. The resulting nine equations can be written as follows:

$$\left(i_{12 \cdot l + k + 9} - i_{w,3 \cdot l + k} + i_{w,3 \cdot l + k - 1}\right)\Big|_{t+h} = 0 \quad , \tag{3.33}$$

$$i_{w,0} = i_{w,9}$$

for $k = 1,2,3$ and $l = 0,1,2$.

A star connection option instead of polygonal would result into:

$$\left(i_{12 \cdot l + k + 9} - i_{w,3 \cdot l + k}\right)\Big|_{t+h} = 0, \tag{3.34}$$

for $k = 1,2,3$ and $l = 0,1,2$.

In addition to these twenty Kirchhoff current equations twenty-six Kirchhoff voltage equations have to be set up in order to complete the system. Three types of loops are chosen here. First there are nine loops directly through the network from earth potential via a grounding resistor, the RL-connector and the line-to-ground voltage back to earth potential.

$$\left(-u_{12 \cdot l + k} - u_{12 \cdot l + k + 9} - u_{A,3 \cdot l + k}\right)\Big|_{t+h} = 0, \tag{3.35}$$

for $k = 1,2,3$ and $l = 0,1,2$.

Furthermore, eight loops are taken through a grounding resistor, the diodes with even branch numbers in immediate neighborhood and via the following RL-connector and the terminal voltage back to ground:

$$\left(-u_{12 \cdot l + k} + u_{12 \cdot l + 2 \cdot (k+1)} - u_{12 \cdot l + 2 \cdot (k+2)} - u_{12 \cdot (l+1) + k - 2} - u_{A,3 \cdot l + k + 1}\right)\Big|_{t+h} = 0 \tag{3.36}$$

for $k = 1,2$ and $l = 0,1,2$.

$$\left(-u_{12 \cdot l + 3} + u_{12 \cdot l + 8} - u_{12 \cdot (l+1) + 4} - u_{12 \cdot (l+2) - 2} - u_{A,3 \cdot l + 4}\right)\Big|_{t+h} = 0,$$

for $k = 3$ and $l = 0,1$.

The remaining 9 loops lead via mere diode paths and the load:

$$\left(u_{12 \cdot l + 2 \cdot (k+1)} + u_{12 \cdot l + 2 \cdot k + 1} + u_{37}\right)\Big|_{t+h} = 0 \tag{3.37}$$

for $k = 1,2,3$ and $l = 0,1,2$.

In comparison with this integrated network solution of the FD calculation an iterative solution is described in [3.25], where the currents in each winding are estimated, before the field in the machine is solved. This approach is only feasible for machines with concentrated windings.

A verification of the mathematical model with measurements is realized for a brushless exciter acc. fig. 3.6. Extensive investigations have been done on this 8-pole brushless exciter with rotating armature. The diode currents have been measured by Rogowski coils. The signals were transmitted by telemetry.

3.2 Non-linear Transient Time-Stepping Numerical Field Calculation

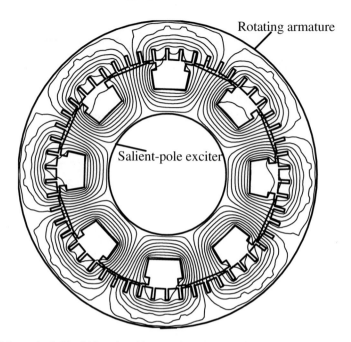

Fig. 3.6 Magnetic field within a brushless exciter during steady state operation coupled to the turbo-generator field winding

The exciter has been measured on an exciter test bench. Therefore the rectified current has been fed in a 0.1 Ω resistor standing for the main turbo-generator field winding. This has been considered in the calculation in the same way. After the steady state is reached the current form is periodic. The calculation results are compared for one period with the measured diode currents in fig. 3.7.

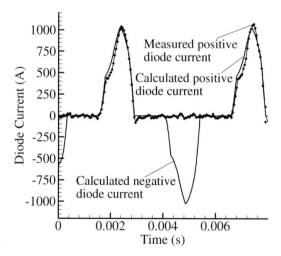

Fig. 3.7 Comparison between measured and calculated diode currents

The agreement between the measured and calculated diode current according to fig. 3.7 is excellent. Both, the width of the current wave as well as the amplitude fit well. During the telemetry measurement on the rotating armature the calculation even revealed an unforeseen filter, which has been avoided then for the following test series. Not only diode currents have been measured by Rogowski coils and telemetry, but the phase currents as well. The phase currents are shown in fig.3.8. Again the calculated phase current agrees well with the measured one.

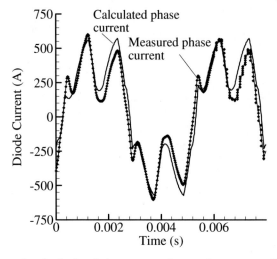

Fig. 3.8 Measured and calculated phase current for a turbo-generator field current of 1500 A and a field voltage of 150 V

Chapter 4
Additional Losses Due to Higher Voltage Harmonics

4.1 Overview of Converter Dependent Loss Components

Loss and efficiency determination are described in IEC 60034-2. These standardized procedures are inevitable in order to derive comparable results between different motor manufacturers and applications. The losses which will occur in reality during converter operation will be different. Table 4.1 gives an overview of the different types of losses acc. IEC 60034-2, [4.1].

Table 4.1 Different types of losses within electrical machines acc. IEC 60034-2

Kind of Loss	Test	Equations (4.1-4.5)	Remarks
Stator d.c. copper loss $P_{cu1,dc}$	d.c. resistance measurement between two terminals $R_{1,dc,n}$, Measurement of rated current I_r	$P_{cu1,dc} = \dfrac{3}{2} R_{1,dc,r} \cdot I_r^2$ $R_{1,dc,r} = R_{1,dc,m} \dfrac{\vartheta_r + k}{\vartheta_m + k}$, with $k = 235K$ (Cu)	No eddy currents considered
Iron losses P_{Fe}	No-load test	$P_{Fe} = P_{1,0} - P_{cu1,0,dc} - P_{Reib}$	Higher harmonic no-load losses are included in P_{Fe}
Friction losses P_{reib}	No-load test	Constant part of no-load test	Bearing losses depend upon running time and motor temperature
Rotor losses	Load test	$P_{cu2} = s \cdot P_\delta$ $= s \cdot (P_{1,r} - P_{cu1,r,dc} - P_{Fe})$	Additional losses are not included
Additional losses	No test	$P_{ad} = 0.5\% \cdot P_r$	The estimation is very rough

This loss structure has been derived based on a sinusoidal machine supply. Basic principles are the same in converter driven machines. Indeed it is important to investigate those components, which are gathered in the component additional losses more in detail. Additional losses do occur in the following machine parts:

O. Drubel: *Converter Appl. & their Influence on Large Electr. Mach.*, LNEE 232, pp. 45–67.
DOI: 10.1007/978-3-642-36282-8_4   © Springer-Verlag Berlin Heidelberg 2013

- Stator winding
- Rotor winding
- Stator core losses
- Rotor core losses
- Losses within housings and other design structural components

4.2 Eddy Current Distribution within the Stator and Rotor Winding

In order to determine the stator copper losses the stator currents have to be known. Converter operation is a steady state operation regarding the losses. The stator currents can be Fourier analyzed and each individual component can be applied to the stator winding loss calculation separately or they can be directly determined from the voltage-current characteristic. The overall losses are determined by the summation of all harmonic losses or by the integration of the loss characteristic in time. The frequency dependent loss calculation needs to consider the eddy current effects within the winding. The eddy current effect consists of different parts. Circulating currents may be induced in parallel connected coils or parallel connected turns. Even if all turns have the same current, eddy currents are induced. A transient solution is given in fig. 4.1

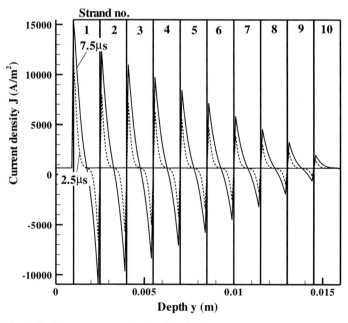

Fig. 4.1 Typical eddy current density distribution in a bar with 10 strands and a time constant of $10\mu s$

4.2 Eddy Current Distribution within the Stator and Rotor Winding

Fig. 4.1 had been calculated by FE-time stepping for an exponential current distribution, see [4.2]. In case of sinusoidal current waveforms the principal distribution of current density is similar, but the course in time is different. The individual copper losses for sinusoidal distribution can be calculated acc. equation 4.6-4.14:

$$P_{vw} = nR_{n0}I^2\left[\varsigma(\beta) + \frac{n^2-1}{3}\psi(\beta)\right],\tag{4.6}$$

$$\varsigma(\beta) = \beta\frac{\sinh(2\beta)+\sin(2\beta)}{\cosh(2\beta)-\cos(2\beta)},\tag{4.7}$$

$$\psi(\beta) = 2\beta\frac{\sinh(\beta)-\sin(\beta)}{\cosh(\beta)+\cos(\beta)},\tag{4.8}$$

$$\beta = \alpha \cdot h,\tag{4.9}$$

$$\alpha = \sqrt{\frac{\omega\chi\mu_0\mu_r}{2}\frac{b_l}{b_n}},\tag{4.10}$$

$$L_{n1} = L_{n10}\left[\frac{\varsigma'(\beta)}{n^2} + \frac{n^2-1}{n^2}\psi'(\beta)\right],\tag{4.11}$$

$$\varsigma'(\beta) = \frac{3(\sinh(2\beta)-\sin(2\beta))}{2\beta(\cosh(2\beta)-\cos(2\beta))},\tag{4.12}$$

$$\psi'(\beta) = \frac{\sinh(\beta)+\sin(\beta)}{\beta(\cosh(\beta)+\cos(\beta))}.\tag{4.13}$$

with b_l strand width, b_n slot width, h strand height, n complete number of individual strands in a slot .

In slots with two winding layers with a phase shift φ_{ph} the equation 4.6 will be modified acc. equation 4.14:

$$P_{vw} = nR_{n0}I^2\left[\varsigma(\beta) + \left(\frac{5n^2-8}{24} + \frac{n^2}{8}\cos(\varphi_{ph})\right)\psi(\beta)\right] = nR_{n0}I^2k_{rtl}.\tag{4.14}$$

An equal current is assumed in any strand within equation 4.6-4.14. It is often the case that the strands are split in two or more for instant z parallel connected strands. These strands are connected in parallel within one stator coil and short circuited at the coil ends. Due to the slightly different position within the stator slot the slot flux linkage is different and circulating eddy currents occur. Therefore the currents in these two strands are slightly different. In a normal two layer winding the strands change their order towards the air-gap after every slot-length due to the way the coil is placed in the slots, [4.3]. The difference in flux linkage is reduced, but still exists, due to the difference between upper and lower layer. The circulating currents cause an increase in losses according equation 4.15-4.17.

$$\alpha^* = \sqrt{\frac{\omega \chi \mu_0 \mu_r}{2} \frac{b_l}{b_n}} \sqrt{\frac{l_{Fe}}{l_{ges}}}, \tag{4.15}$$

$$\beta^* = \alpha^* \cdot z \cdot h, \tag{4.16}$$

$$k_{rut} = \varsigma(\beta^*) + \frac{\left(\dfrac{n}{z}\right)^2 - 4}{16} \psi(\beta^*). \tag{4.17}$$

The relation l_{Fe}/l_{ges} considers the different flux linkage within the end-winding region, assuming that no linkage occurs in this area. This simplification is of cause wrong. In the end winding region also a different flux linkage does occur. The linkage can be approximated by a Biot Savart calculation, see [4.4]. The winding within large 2-pole turbo-generators consists of Roebeled strands in order to minimize the circulating current between parallel connected strands. It is targeted to compensate the different flux linkage of the individual strands completely. In order to reach a proper compensation, which takes the influence of the end winding into account, the bars are not Roebeled by 360° but a little bit less. The difference is about 10% which means, that the difference in flux linkage in the end winding region is about 10% of the linkage within the slot area. The effect is smaller in machines with higher pole numbers. In nearly completely compensated Roebel bars the end winding area is quite important. A loss margin of 10% should be considered for the circulating current losses, if the bars are not Roebeled in the end winding.

The overall loss increase, which is caused by eddy currents in the stator winding, is given in equation 4.18:

$$k_{rge} = k_{rtl} \cdot k_{rut}, \tag{4.18}$$

The same equations could be applied to a squirrel cage induction machine as well for n=1, see equation 4.19.

$$k_{rges,rotor} = \varsigma(\beta) = \beta \frac{\sinh(2\beta) + \sin(2\beta)}{\cosh(2\beta) - \cos(2\beta)}. \tag{4.19}$$

The rotor conductor dimensions often do not fit at all with the shape of a stator strand. It may be better to determine the loss factor numerical by finite element calculations. The loss increase has been calculated for different slot shapes, fig. 4.2, by numerical field calculation. The results are given in fig. 4.3. The totally different bars fulfill their tasks for a frequency below 100 Hz according to the standard machine theory. Hardly any eddy current effects occur below 1 Hz, which is the current frequency of the fundamental at rated operation, strong eddy current effects for frequencies up to 50 Hz help to increase the motor torque during direct on line starts.

An interesting result can be seen in the frequency range, which is important for the evaluation of additional losses due to higher converter harmonics. The losses will increase proportional to $f^{2/3}$ for frequencies above 1000 Hz. The rotor current amplitude has been kept constant for all frequencies f. Some more information about the rotor loss calculation may be found in [4.16, 4.17].

4.2 Eddy Current Distribution within the Stator and Rotor Winding

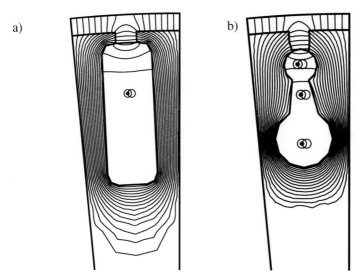

Fig. 4.2 Field distribution at 50Hz for a squirrel cage slot, a) deep bar and b) double cage bar

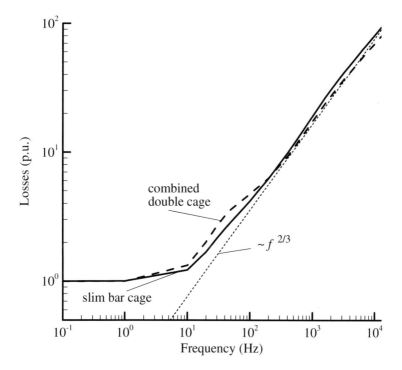

Fig. 4.3 Loss increase vs. frequency for a deep bar and double cage bar at constant rotor current rms, related to an eddy current free loss amount for the same copper area

4.3 Additional Iron Losses within the Stator and Rotor Lamination

Additional losses within the lamination occur beside additional losses within the copper or aluminium windings. The additional iron losses can be calculated in different ways. Two main topics are always within the focus of investigations:

- the influence of a non sinusoidal flux-distribution has to be analysed,
- the time course of induction has to be determined.

Different publications [4.5, 4.6] handle with the first topic. The focus of measurements in [4.5] is based on a sinusoidal supply for two different frequencies, which are superposed in different ways, a trapezoidal characteristic is applied in [4.6]. A realistic approach seems to be the summation of the individual induction harmonic losses and to allow for a margin of 10 % for the loss increase due to their interaction. Numerical field calculation with time stepping could be the answer of the second question. This approach would be theoretically exact, because it could be based on the induction time course within the individual field elements and their individual fourier-analysis. Practically this answer would be too time intensive. A simpler approach is based on the terminal voltage. The integration of the voltage with time is equivalent with the total winding flux linkage, if any winding resistance is neglected. This flux linkage would be fourier-analysed to get the individual harmonic component in flux Ψ_v. If a similar flux distribution would be assumed for any higher harmonic as for the fundamental, the additional iron losses in laminations can be based on the iron losses of the fundamental. Equation 4.20 can be applied for iron losses at different frequencies and inductions, see [4.7]:

$$P_{\text{Fe},v} = C_{Fe} \cdot f_v^{1.36} \cdot B_v^{1.71} \tag{4.20}$$

Even though [4.7] proposed the validity of the equation 4.20 up to 500 Hz, table 4.2 shows that excellent results can be derived for 1000 Hz as well.

Table 4.2 Verification of equation 4.20 with measurements acc. [4.5]

	50Hz – measured [4.5]	1000Hz – measured [4.5]	1000Hz – eq. acc. 4.20
Losses at 1.7 T	12.9W	-	
Losses at 0.15 T	-	11.6W	11.9W
Losses at 0.20 T	-	19.9W	19.5W

The converter harmonics can be taken into account by an additional loss factor k_{Fe}, acc. equation 4.21. The interaction of non-sinusoidal inductions is represented by the interaction factor k_{inter}. This factor is in the range of 10 %.

$$P_{Fe} = \sum_v P_{\text{Fe},v} = P_{\text{Fe},1} \cdot \left(1 + k_{\text{Fe}} \cdot k_{\text{inter}}\right) = P_{\text{Fe},1} \cdot \left(1 + k_{\text{inter}} \cdot \underbrace{\sum_v v^{1.36} \cdot \left(\frac{\Psi_v}{\Psi_1}\right)^{1.71}}_{k_{\text{Fe}}}\right). \tag{4.21}$$

4.3 Additional Iron Losses within the Stator and Rotor Lamination

The flux distribution for the fundamental and the higher harmonics are not the same. The fundamental consists of "main flux" and "leakage flux", the higher harmonics are damped by rotor currents. Mainly leakage flux exists, see fig. 4.4 b).

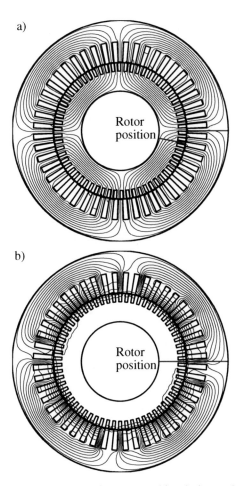

Fig. 4.4: a) Flux distribution in an asynchronous machine during no load operation; b) flux distribution with locked rotor

The field distribution and the iron losses in stator and rotor have been calculated for the numerical field calculations fig. 4.4 a, b. The losses are given in table 4.3.

Table 4.3 Iron losses during no load operation and with locked rotor current for the same terminal voltage

	No load operation	Locked rotor
Stator iron losses at 50Hz	11800W	13700W
Relationship locked rotor losses / no load losses		1.15

The non-linear calculated losses during no load operation are about 15% smaller than for the locked rotor, if the voltage is kept the same. The additional losses are especially generated in the stator teeth.

4.4 Eddy Current Losses within Massive Magnetic Material

Additional losses do occur on any conducting design parts within electrical machines, which are exposed to magnetic flux. These parts can be surfaces of the rotor in synchronous machines, press-plates, iron housings, support rings or end bells. Most of these parts consist of iron. Due to the non-linearity of iron special analytical models from Gibbs and Oberretl [4.8, 4.9] had been derived in the past which are compared with numerical field calculation in [4.15]. The analytical methods have been developed mainly for flux ripple on salient pole synchronous machines, but they can be applied for all other kinds of machines and for converter fed ones as well. The methods show their advantages especially at design areas with massive iron material. Here eddy-current problems often strongly dependent on the saturation of the iron. In contrast to numerical methods, which can take saturation into account, analytical methods require the magnetic properties to be linearised by using some form of equivalent permeability. In addition, an analytical method has to simplify the geometry of the real machine and use a mathematical model such as that shown in fig. 4.5.

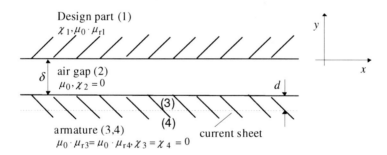

Fig. 4.5 Mathematical model for flux pulsation calculation, with air-gap δ, conductivity χ and permeability μ, see [4.10]

This model represents the flux pulsation by a current sheet adjacent to an airgap with a smooth surface. The 'pole pitch' of the current sheet is assigned with λ. The magnitude of the current is adjusted to the first harmonic of the flux pulsation at the design part surface for frequency $\omega = 0$, i.e. in the absence of eddy currents, to a level that it is the same as for the real machine. Gibbs [4.9] uses a harmonic factor to take the additional loss due to the higher harmonics into account. The influence of the mean flux density is neglected.

4.4 Eddy Current Losses within Massive Magnetic Material

A finite element calculation is applied to the same mathematical model as a comparison. It does not use a time-stepping method, but solves the Helmholtz equation 4.22 with non-linear magnetic parameters in each element:

$$\text{rot}\left(\frac{1}{\mu}\text{rot}\begin{pmatrix} 0 \\ 0 \\ \underline{A}_z \end{pmatrix}\right) = \underline{J}_z - j\omega_{\text{el}}\chi\underline{A}_z, \tag{4.22}$$

where $\underline{A}_z$ is the complex axial component of the magnetic vector potential and $\underline{J}_z$ the impressed current density.

The permeability in each element is independent of time, but differs from element to element and depends on the flux amplitude in this element. The magnitude of the vector potential varies sinusoidally with time. The conductivity χ is the same throughout the design part material.

A finite difference time stepping calculation is applied as a third method, which allows for saturation as a function of time and takes the moving boundaries into account [4.11]. Here the influence of the mean flux density is investigated. The method uses two meshes; one in the stator and one in the rotor. Rotation is made possible by an adjustment between the two meshes at an interface in the air-gap. The comparison between the four methods, which means the two numerical and the two analytical ones, is done for two different magnetization curves of rotor steel. A wide range of pulsation amplitudes and frequencies is considered. First the analytical methods are presented in detail. The methods of Gibbs and Oberretl are well known and the theory can be found in literature [4.8, 4.9]. Gibbs' theory assumes the field to be one dimensional. This means the field is tangential within the pole face. The field calculated with finite elements is shown in fig. 4.9, where it can indeed be seen to be mainly tangential in the pole. It does have a radial component mainly in those regions where the field is small. Gibbs bases his estimation of permeability on the maximum value of this tangential component, but ignores the reaction effect of the eddy currents in reducing the applied ripple field. In the following, this basic one-dimensional method is used, but, in place of the harmonic factor in [4.9], which is introduced to allow for harmonic losses, the higher harmonics are taken into account by the more accurate factor in [4.8], given below in equation 4.23 as k_L. With these assumptions the loss per unit surface area A_S, in SI units are determined:

$$\frac{P_{\text{ri}}}{A_S} = \frac{\lambda^{-1}\hat{B}_{0,\delta} \cdot f_{\text{el}}}{4} \cdot \hat{H}_\theta \cdot \sqrt{2} \cdot k_L, \tag{4.23}$$

with the maximum value of the tangential θ-component of the field strength:

$$\hat{H}_\theta\sqrt{\mu_0 \cdot \mu_r} = \frac{\sqrt{f_{\text{el}} \cdot \chi} \cdot \lambda^{-1}\hat{B}_{0,\delta}}{\sqrt{2\pi}}, \tag{4.24}$$

where $^1\hat{B}_{0,\delta} = \alpha_1 \cdot B_{m,\delta}$.

The factor α_1 gives the relation between the first harmonic of the radial flux density oscillation at the pole surface and the mean value of the flux density B_m over one slot pitch. α_1 is given in fig.4.6 as a function of slot width s, slot-pitch λ and air-gap length δ. It has been obtained both in [4.8] and [4.9] using the technique of conformal transformation, where the surfaces are assumed to be infinitely permeable.

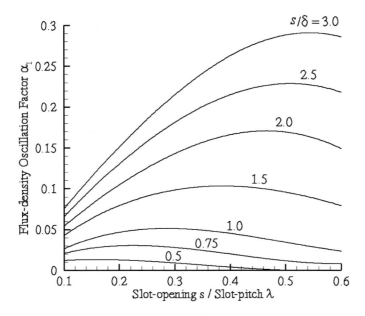

Fig. 4.6 Flux oscillation factor α_1 for different machine geometries, acc. [4.8]

The tangential field can be determined from equation 4.24 using fig. 4.7 to provide the relative permeability. The magnetization curves for the two steels, which have been considered, have been presented in the unusual form shown in fig. 4.7, because the product on the left-hand side of equation 4.24 is fixed by the operating condition and is essentially independent of the permeability. The theory in [4.8] is more accurate by the application of the two-dimensional solution of the field problem according fig. 4.5, which had been outlined by Lawrenson [4.10]. However, Lawrenson did not consider the problem of selecting a suitable permeability. Following the method in [4.9], a field-dependent permeability is introduced, but Oberretl found experimentally that better results are obtained by using only 85% of the maximum tangential field rather than the maximum value.

4.4 Eddy Current Losses within Massive Magnetic Material

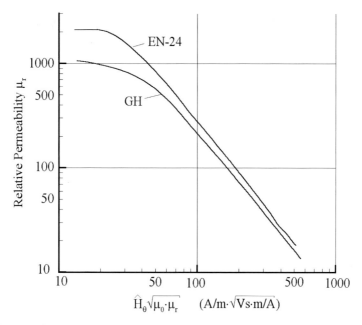

Fig. 4.7 Determination of a substitute permeability μ_r for two different magnetization curves; EN-24 is a curve acc. [4.12] and GH is a curve acc. [4.11]

The theory includes the damping effect of the eddy currents and makes allowance for the effect of saturation in the armature teeth tips on the higher pulsation harmonics. The losses are given by equation 4.25:

$$\frac{P_{ri}}{A_S} = \frac{\hat{B}_{d,\delta}^2 \cdot \lambda^2 \cdot f_{el}^2 \cdot \chi}{4 \cdot \text{Re}(\gamma)} \cdot k_L, \qquad (4.25)$$

where the damping effect of eddy currents has been included by modifying the original first harmonic component of the radial air-gap tooth-ripple flux density at the pole surface ${}^1\hat{B}_{0,\delta}$ according to equation 4.26, see [4.8, 4.15], indices according fig.4.5:

$${}^1\hat{B}_{d,\delta} = {}^1\hat{B}_{0,\delta} \cdot \left(\frac{(1-\mu_{r3}) \cdot e^{-\frac{2\pi}{\lambda}\delta} \cdot (\mu_{r1}-1) + (1+\mu_{r3}) \cdot e^{\frac{2\pi}{\lambda}\delta} \cdot (\mu_{r1}+1)}{(1-\mu_{r3}) \cdot e^{-\frac{2\pi}{\lambda}\delta} \cdot \left(\mu_{r1} - \frac{\lambda}{2\pi}\gamma\right) + (1+\mu_{r3}) \cdot e^{\frac{2\pi}{\lambda}\delta} \cdot \left(\mu_{r1} + \frac{\lambda}{2\pi}\gamma\right)} \right). \qquad (4.26)$$

The skin depth parameter γ is given by equation 4.27:

$$\gamma = \sqrt{\left(\frac{2\pi}{\lambda}\right)^2 - j\chi \cdot 2\pi \cdot f_{el} \cdot \mu_0 \cdot \mu_{r,ob}}, \qquad (4.27)$$

The harmonic factor k_L includes an allowance for the presence of tooth-ripple harmonics:

$$k_L = \left(\frac{B_{0,m,\delta}}{{}^1 B_{0,\delta}}\right)^2 \cdot \sum_{\nu=0}^{\infty} \frac{1}{\sqrt{n}} \left(\frac{{}^\nu B_{0,\delta}}{B_{0,m,\delta}}\right)^2, \qquad (4.28)$$

where ν denotes the harmonic number, ${}^\nu \hat{B}_{0,\delta}$ the peak value of the ν-th harmonic flux density at the pole surface and $\hat{B}_{0,m,\delta}$ the mean flux density in the air-gap over one slot pitch.

Values of k_L used for both analytical theories are given in fig. 4.8.

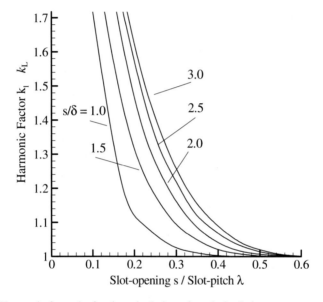

Fig. 4.8 Harmonic factor k_L for the calculation of tooth ripple losses

The equivalent permeability of the pole face $\mu_{r,Ob}$ is determined in equation 4.29:

$$\sqrt{\mu_0 \mu_{r,Ob}} \hat{H}_\theta = \frac{\lambda}{2\pi} \cdot 0.85 \cdot \hat{B}_{d,\delta} \cdot |\eta| \cdot \frac{1}{\sqrt{\mu_0 \mu_{r,Ob}}}. \qquad (4.29)$$

Equation 4.29 is applied iterative in conjunction with fig. 4.7. The process is an iterative one, but converges after a few steps. Even though the analytical methods give relatively fast results, once they are implemented they should be proven by numerical methods. A finite element method [4.13] solves Helmholtz equation 4.22, and has been applied to the mathematical model of fig. 4.5. The losses have been calculated with a permeability which is constant in time. The tooth-ripple field at the pole surface has been generated by a thin current-carrying layer on the

4.4 Eddy Current Losses within Massive Magnetic Material

surface of the armature consisting of 10 segments in the circumferential direction, each with the same magnitude of current density but a changing phase to model relative movement with respect to the pole. This current distribution is shown in fig. 4.9. Adaptive mesh generation with automatic mesh refinement based on the flux density has been used. The boundaries are periodic in the traveling direction and the vector potential is set to zero at the other boundaries, where the field is negligible. The non-linear version of the solution determines a different permeability in each element of the pole according to the local flux density, and is based on the iterative Newton-Raphson algorithm. Before the iteration process commences a magnetization curve is prepared based on the assumption of sinusoidal field strength, H, and therefore non-sinusoidal flux density, B. The curve used gives the rms value of the first harmonic of flux density against the rms value of field strength. This step is necessary because Helmholtz' equation represents the field in complex form.

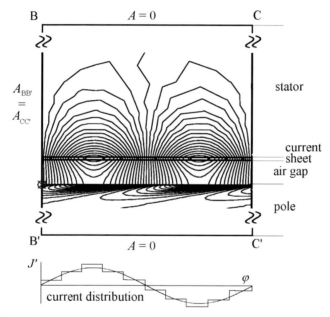

Fig. 4.9 Flux density according to the model of fig. 4.5; only the interesting sections of the field problem are shown; there are some field lines shown which close outside the region, which is illustrated

The flux density obtained by the finite element method is shown in fig. 4.9 for the non-linear version of the problem. As mentioned earlier, the field inside the pole is mainly tangential to the surface for the dimensions and frequencies considered. The field of the non-linear calculation penetrates further into the pole than for the linear calculation using the equivalent permeability according to [4.8]. This can be seen in fig. 4.10 a), where the amplitude of the magnetic flux density is shown for both cases and in fig 4.10 b), where the eddy current density is given.

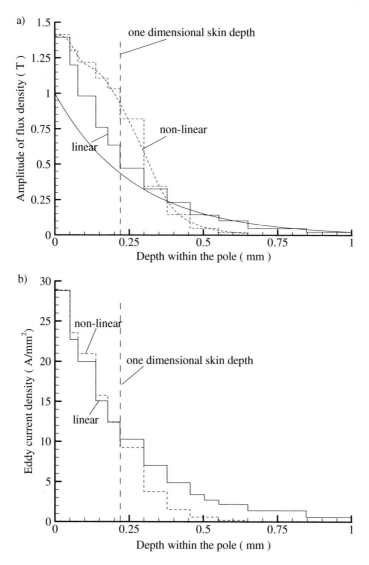

Fig. 4.10 Amplitude of flux density a) and eddy current density b) the field problem of fig. 4.9 for $f_{el} = 2800$ Hz and $^1\hat{B}_{d,\delta} = 0.28$ T

The polynomial and exponential curves are derived from the discrete values of flux density in each element. The linear solution decays exponentially with depth, and it is found that the skin depth of this two-dimensional field is almost exactly equal to that for the one-dimensional case. The tooth ripple loss per unit surface is 13.0 kW for the linear case and 12.6 kW for the non-linear case.

The harmonic factors of the slot arrangement have been compared with the linear analytical solution of Lawrenson [4.10] by a second numerical model of the

4.4 Eddy Current Losses within Massive Magnetic Material

slot. The error with a finite element mesh of 3000 elements was found to be smaller than 0.7%. This error depends on the discretization of the mesh and the current sheet. The assumptions, which have been applied in the previous methods, are overcome by a numerical time stepping method with moving boundaries. The analytical and finite element methods treat the losses, which are caused by the fundamental and higher harmonics, separately because a complex vector potential solution is employed. However, with time-stepping and moving boundaries no such separation is necessary. The losses are caused by the relative movement of the toothed structure as a whole. Although hysteresis and material anisotropy are still neglected, the particular advantage of the applied finite difference program is the possibility of allowing movement between the rotor and stator. The real geometry of the machine is considered and instead of a substitute current sheet, the eddy currents are induced by movement relative to the armature teeth.

Relative motion v between two parts such as the rotor and stator of a cylindrical machine can be taken into account in different ways. The basis for any calculation is Faraday's law according equation 4.30:

$$\text{rot}\left(\vec{E} - \vec{v} \times \vec{B}\right) = -\frac{d\vec{B}}{dt} \,. \tag{4.30}$$

If a single coordinate system is set up and fixed with respect to the stator, the rotor moves relative to this system and the complete equation has to be considered. Equation 4.30 can be simplified according equation 4.31, if the rotor and stator are divided into two adjacent coordinate systems:

$$\text{rot}\vec{E} = -\frac{d\vec{B}}{dt} \,. \tag{4.31}$$

The influence of the movement is taken into account by the boundary between the stator and rotor field regions in the air-gap. As a result, the eddy-current diffusion equation 4.32 with an external field E_a, can be applied in the rotor:

$$\nabla^2 A_z = \chi \mu \cdot \left(\frac{\partial A_z}{\partial t} + E_a\right). \tag{4.32}$$

In both regions the space is discretized by finite difference meshes, and the time stepping handled by the Crank-Nicolson algorithm. Thus, the original differential equation is transformed to a set of algebraic equations that are solved directly to obtain the vector potential in the rotor and stator coordinate systems for each time step. The term E_a in equation 4.32 allows the influence of impressed voltages and three-dimensional end effects, which could arise for instance by the approximate modeling of the end rings in an asynchronous machine, to be taken into account. A sample result is shown in fig. 4.11, where each contour indicates a constant value of current density at one instant in time.

60   4 Additional Losses Due to Higher Voltage Harmonics

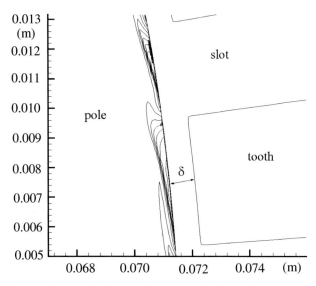

Fig. 4.11 Current density distribution for the tooth ripple phenomenon; the effect of the mean flux density and moving boundaries are taken into account

The finite difference solution takes the mean flux density into account, so that the flux density never becomes zero, see fig. 4.12.

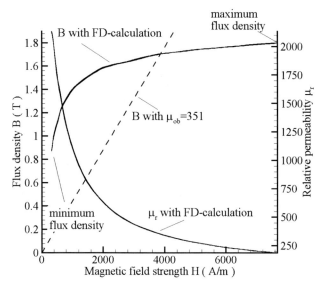

Fig. 4.12 Flux density during one ripple cycle, with $f_{el}= 2000$ Hz and $^1\hat{B}_{d,\delta} = 0.28$ T

4.4 Eddy Current Losses within Massive Magnetic Material

Fig. 4.12 shows a typical excursion of the flux density at one point in the rotor over one cycle of the tooth-ripple pulsation. In addition to the non-linear calculation the linear *B-H* curve with an equivalent permeability according to [4.8] is shown. This equivalent permeability is determined from the a.c. field alone. An equivalent permeability that takes the d.c. field into account will have a smaller value. Additional losses in mechanical structure parts or massive iron surfaces depend strongly upon the right consideration of material saturation. Even though the presented time stepping method takes these saturation effects into account correctly, its application cause a lot of efforts and time consumption. So it is often not applicable in real life. A wide range of application can be considered if mayor basics are prepared. The tooth-ripple losses depend on the degree of saturation within the iron. On the other hand, if the field can be regarded as one-dimensional equations 4.23 and 4.24 show that the ripple loss per unit surface area is proportional according to equation 4.33:

$$\frac{P_{\mathrm{ri}}}{A_{\mathrm{S}}} \propto f_{\mathrm{el}}^{1.5} \cdot \lambda^{2.1} \hat{B}_{\mathrm{d,\delta}}^2 \cdot k_{\mathrm{L}} \cdot \sqrt{\chi} \, . \tag{4.33}$$

The tangential field amplitude and the saturation determinate the flux density. The effect is represented by equation 4.34:

$$\hat{H}_{\theta} \sqrt{\mu_0 \cdot \mu_{\mathrm{r}}} \propto \underbrace{\sqrt{f_{\mathrm{el}} \cdot \chi} \cdot \lambda^{1} \hat{B}_{\mathrm{d,\delta}}}_{S_{\mathrm{dr}}} \, . \tag{4.34}$$

The term on the right-hand side of 4.34 is called the saturation factor S_{dr}. This saturation factor is listed in table 4.4 for four industrial synchronous machines.

Table 4.4 Saturation factor and its application to four different industrial synchronous machines. The losses have been calculated according to Oberretl's method for two different conductivities, $\chi = 4 \cdot 10^6$ S/m, case A, and $\chi = 8 \cdot 10^6$ S/m, case B

rating (MVA)	measured losses (kW)	pole shoe loss / no-load iron loss	saturation factor $\left(\sqrt{\mathrm{Hz\,S/m}}\ \mathrm{T\,m}\right)$		calculated losses (kW)	
			A	B	A	B
16.6	5.2	8.0 %	160	219	6.3	10.2
13.1	10.3	19.0 %	169	231	6.2	10.6
7.4	11.7	36.4 %	215	293	9.5	16.7
6.3	12.6	33.9 %	250	338	12.1	20.2

A comparison of the methods for a wide range of parameters is possible by a plot of the loss per unit surface area divided by the right-hand side of equation 4.33 against the saturation factor, see fig. 4.13:

62   4 Additional Losses Due to Higher Voltage Harmonics

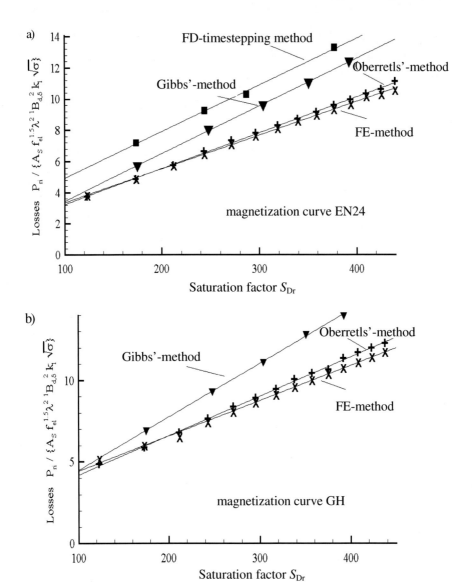

Fig. 4.13 a) Power losses for different degrees of saturation and calculation methods, magnetization curve EN24, conductivity $\chi = 6.89 \cdot 10^6$ S/m; b) Power losses for different degrees of saturation according to all four calculation methods, magnetization curve GH, conductivity $\chi = 6.89 \cdot 10^6$ S/m

The analytical methods are compared with the numerical calculations, for the two pole steels, which have been considered in fig. 4.13.

4.5 Loss Measurement within Magnetic Material

Deviations of less than 6 % are found between the analytical method according to [4.8] and the finite element solution when applied to the same model. In contrast, Gibbs' method gives losses that are up to 25 % higher. Using different magnetization curves for the same calculation method causes a difference of up to 14 %.

The non-linear finite difference time-stepping calculation has only been applied for one magnetization curve, see fig. 4.13 a), due to the immense computing effort, which is involved. The losses are up to 40 % higher than with the conventional method in [4.8]. A simulation with the other magnetization curve would be expected to give similar results. However, a linear finite difference calculation yields losses that are only about 3% more than those from Oberretl's method. The substantial increase in the non-linear version must be caused by the proper consideration of the mean flux density and the difference in handling the higher harmonics, which results from the solution of the complete non-linear problem for the geometry in a single computation. Differences due to discretisation errors are only in the range of some few percent.

4.5 Loss Measurement within Magnetic Material

The direct measurement of additional losses is hardly possible in a machine as a whole. Measurements on special models or the direct measurement of the electrical field or the magnetic field strength may open solutions for a direct measurement. This is shown in the example of eddy current losses in non-linear material. A model has been built in which tooth-ripple losses are virtually the only iron loss. Measurements were conducted with a torque meter and a Poynting-vector probe, see [4.14]. The probe, shown in fig. 4.14, consists of three coils. One coil is designed to measure the electric field and the other to determine the tangential component of the magnetic field strength. A third coil is needed to compensate for the strong radial component of air-gap flux that links the narrow circumferential span of the second coil. The signals from the E - and H - coils are shown in fig. 4.15. The electric field is directly proportional to the radial air-gap flux density. The compensated signal of the H-coil depends mainly upon the eddy currents in the pole iron. The losses are calculated according to the Poynting vector method, if both signals are Fourier space transformed and their corresponding amplitudes multiplied.

Although the model has very small dimensions compared with a real synchronous generator, it is important to reach a degree of pole surface saturation similar to that in a real generator. This can be done in two ways. The frequency can be increased by the square of the ratio of linear dimensions or the magnitude of the tooth-ripple flux density can be scaled in proportion to this ratio by adjusting the model excitation current. If both adjustments are used as far as possible, only the lower end of the realistic range could be reached on the shown example.

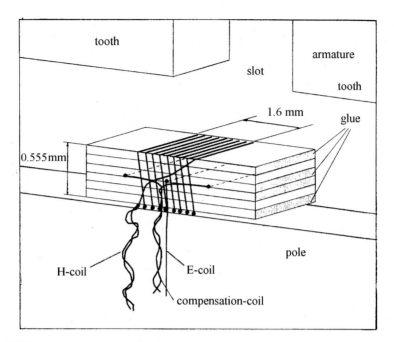

Fig. 4.14 Poynting-vector probe for the measurement of the electric and magnetic field; a compensation coil is used to eliminate the radial flux within the H-coil

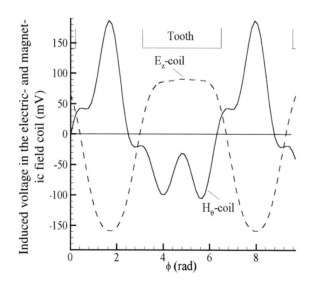

Fig. 4.15 Measured signal of the E-coil and compensated H-coil

4.6 Additional Losses for Different Voltage Source Converter Types

The measurements with the torque meter and Poynting-vector probe are compared with the calculated losses using Oberretl's method, assuming the temperature of the pole iron to be 20 °C, see fig. 4.16. Of course the losses cause an increase of pole temperature, which was unfortunately not measured, but a temperature of 80 °C instead of 20 °C at the pole face increases the rotor resistance by about 40 %. The tangential field strength and the calculated losses decrease with this change in the resistance.

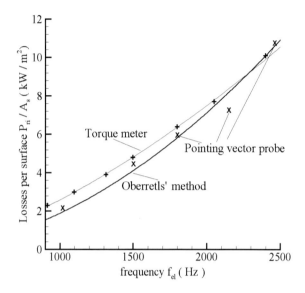

Fig. 4.16 Measured and calculated losses in the model

4.6 Additional Losses for Different Voltage Source Converter Types

Additional losses in a converter driven electrical machine depend strongly on the harmonic content of the imposed voltage. The voltage characteristic is a result of the individual converter type and control strategy. Several converter types have been described in the chapter about typical converter designs for electrical machines. The absolute influence on machine losses will be shown on a medium size asynchronous machine for three voltage source converter types. These types are dominant within industrial drive applications. The converter voltage characteristics have been shown already in the chapter regarding the converter types in detail. They are shown again to allow a direct comparison of the characteristic in fig. 4.17.

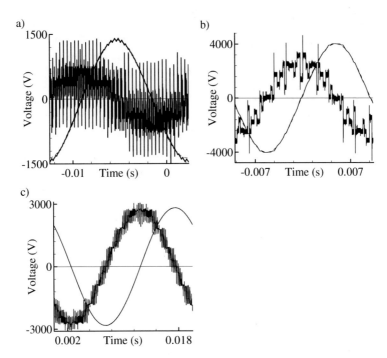

Fig. 4.17 a) Phase to neutral voltage characteristic of a two-level low voltage converter; b) Voltage characteristic of a three-level medium voltage converter; c) Voltage characteristic of a multi-level medium voltage converter with low voltage cells

Each voltage characteristic has been split in its harmonics. Based on the harmonic content additional losses due to converter operation have been calculated within the stator winding, the rotor cage and the stator core. The additional iron losses are given in table 4.4.

Table 4.4 Iron losses for different voltage source converters

	Two-level LV converter	Three-level MV converter	Multi-level MV converter with LV cells
Harmonic iron loss factor k_{fe}	0.27	0.06	0.01
Iron loss factor $(1+k_{inter}\, k_{fe})$	1.30	1.07	1.01

Even though the flux ripple is small in amplitude, relatively high frequencies are reached for the LV converter. The iron losses are 27 % higher than for a sinusoidal supply with the fundamental. The three level MV converter is only 6 % higher in loss than with sinusoidal supply and the MV converter with LV cells only 1 % which can be neglected. The additional iron losses are given together with additional copper losses in fig. 4.18.

4.6 Additional Losses for Different Voltage Source Converter Types

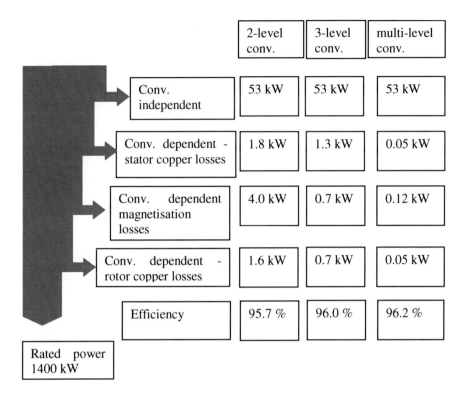

Fig. 4.18 Additional copper and iron losses in a converter driven electrical machine

The influence of higher converter harmonics due to the stator slots has been neglected with respect to the losses. The converter dependent copper losses are relatively high for the two-level converter and can be neglected for the multi-level converter.

Chapter 5
Converter Caused Torque Oscillations

5.1 Interaction between Stator Field Components

Current sheets of the same space harmonics, which impose flux-density components of different phase velocities in the air-gap, cause oscillations in the electrical torque. These may damage the shaft or coupling. The difference in phase velocity causes a time dependent phase shift in flux density components. The torque, which is given by the product of the radial and tangential flux density components, is shown in fig. 5.1. The case with the same space harmonics in both components is given in fig. 5.1 a), b) and with different space harmonics in fig. 5.1 c), d). The integration around the shaft circumferential will be zero independent on the phase shift of the components, if the space harmonics are different, see fig. 5.1 c), d). It will deviate from zero and depend on the phase shift between the flux-density components in case of the same space harmonics, but for instance different frequencies in time, see fig. 5.1 a), b). If the phase shift is constant in time, the flux components will contribute to the average torque. If the phase shift changes in time, the flux components will cause a torque oscillation. Higher frequency harmonics of the same space harmonics may even be induced in case of machines, which are directly operated on the grid [5.1-5.5, 5.12]. In this case they will be called parasitic oscillation torques. Any sinusoidal stator current will induce a rotor current with a frequency according the slip. Higher frequencies within the rotor will be induced as well, due to the space harmonics of the stator field. The rotor cage acts as a winding with space harmonics in the air-gap flux due to the position of rotor bars, which again will induce currents in the stator. Indeed this effect could go on infinitely, but the different frequencies are limited.

O. Drubel: *Converter Appl. & their Influence on Large Electr. Mach.*, LNEE 232, pp. 69–84.
DOI: 10.1007/978-3-642-36282-8_5   © Springer-Verlag Berlin Heidelberg 2013

70 5 Converter Caused Torque Oscillations

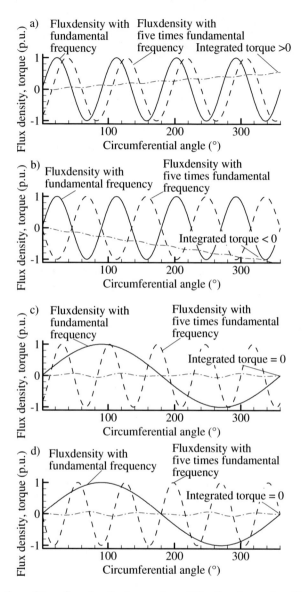

Fig. 5.1 Interaction of two flux density harmonics within the air-gap for different phase shifts between stator and rotor field; a), b) in case of the same number of pole-pairs and c), d) unequal number of pole-pairs

5.2 Calculation of Electromagnetic Torque Oscillations

Table 5.1 gives an overview of the interaction between stator and rotor harmonic generation, more detailed information is available in [5.1 - 5.5].

Table 5.1 Stator and rotor frequencies due to higher harmonics for sinusoidal voltage supply

	Space harmonics	Rotor frequency	Stator frequency
	$^{v}B_1 \propto \exp j\left(\omega_{el,Netz} \cdot t - v \, x_1\right),$	$^{v}\omega_2 = ^{v}s \cdot \omega_{el,Netz}$	$\omega_{el,Netz}$
	$v = p\,(2ma + 1),$ with $a = 0, \pm1, \pm2, \pm3, \dots$	$^{v}s = 1 - \dfrac{v}{p}(1-s)$	$s = \dfrac{\omega_{el,Netz} - p\omega_{mec}}{\omega_{el,Netz}}$
First armature reaction	$^{\mu,v}B_2 \propto \exp j\left(\omega_{v,2} \cdot t - \mu \, x_2\right),$ $\mu = b\,Z_2 + v,$ with $b = 0, \pm1, \pm2, \pm3, \dots$	$^{v}\omega_2 = ^{v}s \cdot \omega_{el,Netz}$	$^{\mu}\omega_1 = ^{\mu}s \cdot \omega_{el,Netz}$ $^{\mu}s = 1 - \dfrac{\mu - v}{p}(1-s)$ $= 1 + (\mu - v)\cdot p\omega_{mec}$
Second armature reaction	$^{\lambda,\mu}B_1 \propto \exp j\left(^{\mu}\omega_1 \cdot t - \lambda \, x_1\right),$ $\lambda = 2mpc + p + b\,Z_2,$ with $c = 0, \pm1, \pm2, \pm3, \dots$	$^{\lambda}\omega_2 = ^{\lambda}s \cdot \omega_{el,Netz}$ $^{\lambda}s = ^{\mu}s - \dfrac{\lambda}{p}(1-s)$	$^{\mu}\omega_1 = ^{\mu}s \cdot \omega_{el,Netz}$
Tertiary armature reaction	$^{\varepsilon,\lambda}B_2 \propto \exp j\left(^{\lambda}\omega_2 \cdot t - \varepsilon \, x_2\right),$ $\varepsilon = mpc + p + d\,Z_2$ with $d = 0, \pm1, \pm2, \pm3, \dots$	$^{\lambda}\omega_1 = ^{\lambda}s \cdot \omega_{el,Netz}$	$^{\varepsilon}\omega_1 = ^{\varepsilon}s \cdot \omega_{el,Netz}$ $^{\varepsilon}s = ^{\lambda}s + \dfrac{\varepsilon}{p}(1-s)$

It is evident, that higher harmonics due to the converter supply may cause a larger amount of stator current harmonics. An additional range of frequencies especially with the pulse frequency of the converter, but also with higher harmonics of the base frequency are generated. An example for the concentration on the main torque components can be found in [5.14]. The higher harmonic stator currents due to the converter operation will create torque oscillations mainly together with the flux distribution of the fundamental. The fundamental frequency causes the flux, whereas the higher converter caused frequencies will act as if the rotor is short circuited.

5.2 Calculation of Electromagnetic Torque Oscillations

Electromagnetic forces within the air-gap of electrical machines can be calculated in different ways. In order to get an impression of the higher harmonics which are induced by the converter a circuit approach with the voltage equations acc. chapter 3 would be sufficient. If a detailed analysis is requested, which considers even the non-linear saturation influence of magnetic slot wedges a numerical calculation, which would be fed by a converter specific voltage characteristic is necessary. This approach will be chosen in the case of noise analysis in chapter 7, where not only the integral of the forces in the air-gap, but the local force distribution is needed. As an alternative a torque calculation process, which is based on the

72 5 Converter Caused Torque Oscillations

superposition of fundamental and harmonic currents, is presented in the actual chapter. A comparison between torque oscillations, which are imposed by a frequency converter, and inherently existing parasitic torques oscillations is given.

In case of a converter the stator voltage consists of a certain spectrum of harmonics. All harmonics, which are not already included within the stator current spectrum, will cause extra torque oscillation frequencies. Either an often sufficient practical approach like in [5.14] is applied or the theoretical steps, which have been described in table 5.1, have to be done for all main voltage harmonics to determine the frequencies within the stator current. This quite theoretical approach is described in the following. If the non-linear characteristic of the stator and rotor iron is neglected, a linear system can be assumed, where the voltage and current harmonics are superposed. The voltage equation within the stator and rotor will be solved for each harmonic. The equation system is well known in literature, [5.3- 5.5].

The shaft torque can be determined based on the calculated current harmonics and field harmonics. The calculation procedure of the torque starts with the determination of the magnetic energy in the machine. The magnetic energy is given according equation 5.1:

$$w_{m} = \frac{1}{2 \cdot \mu_0} \cdot \iiint\limits_{air-gap} \underline{B} \cdot \underline{B}^* dV = \frac{l_{Fe} \cdot \delta \cdot r_i}{2 \cdot \mu_0} \cdot \int\limits_0^{2\pi} \underline{B} \cdot \underline{B}^* d\varphi = \frac{1}{2} \sum_i \psi_i \cdot i_i , \qquad (5.1)$$

with the summation over all machine circuits i.

The torque at the motor shaft is calculated out of the difference from the virtual change in the magnetic energy for constant flux minus the virtual change in the magnetic energy for constant currents with a virtual rotor rotation β_2:

$$M(t, \beta_2) = \frac{\partial w_m}{\partial \beta_2}\bigg|_{\psi=const.} - \frac{\partial w_m}{\partial \beta_2}\bigg|_{i=const.} = \underbrace{\frac{1}{2} \sum_{k=1}^{m} \left(\psi_{1,k} \frac{\partial i_{1,k}}{\partial \beta_2} - i_{1,k} \frac{\partial \psi_{1,k}}{\partial \beta_2} \right)}_{M_1}$$

$$+ \underbrace{\frac{1}{2} \sum_{a=1}^{Z_2} \left(\psi_{2,a} \frac{\partial i_{R,a}}{\partial \beta_2} - i_{2,a} \frac{\partial \psi_{2,a}}{\partial \beta_2} \right)}_{M_2} . \qquad (5.2)$$

Indeed the torque consists of one part M_1, which reveals the change in energy within the stator and of a second part M_2, which depends upon the change in energy within the rotor. A more detailed analysis of equation (5.2) will lead to the torque parts according equation 5.3-5.6 in case of a symmetric asynchronous machine with symmetrical phase shift between the individual stator and rotor currents. The torque calculations are based upon the equations of [5.1, 5.2].

These equations are based on sinusoidal supply, but several harmonics occur due to a converter supply. An extension of the sinusoidal approach can be found as well in [5.6]. Each torque component has to be determined for the higher

5.2 Calculation of Electromagnetic Torque Oscillations

harmonics flux and currents as well. The equations are extended by two further summaions:

$$M_{1,I}(t,\beta_2) = -\frac{m \cdot Z_2}{2} \cdot$$

$$\text{Re}\left\{ \sum_{Um1} \sum_{Um2} \sum_b \sum_{b^"} j(b^" - b) \cdot {}^b\underline{\Psi}_{1,Um1} \cdot {}^{b^"}\underline{I}_{1,Um2} \cdot \exp j\left(\left({}^{b^"}s_{Um2}\omega_{el,Um2} + {}^b s_{Um1}\omega_{el,Um1} \right) t - \left(b^" + b \right) \cdot Z_2\beta_2 \right) \right\}.$$

$$(5.3)$$

The factor Z_2 and $(b^" - b)$ is the result of the derivative of the stator flux linkage against β_2 minus the current derivative. The summation over k stator circuits results in the coefficient of number of phases m. The summation is only not zero, if the following condition is fulfilled:

$$\left(U_{m1} + U_{m2} + \frac{(b^" + b)Z_2}{p} \right) \frac{1}{m} = mg, \qquad g = 0, \pm 1, \pm 2, \dots \quad .$$

The following term is necessary due to the complex approach:

$$M_{1,II}(t,\beta_2) = -\frac{m \cdot Z_2}{2} \cdot$$

$$\text{Re}\left\{ \sum_{Um1} \sum_{Um2} \sum_b \sum_{b^"} j(b^" + b) \cdot {}^b\underline{\Psi}_{1,Um1}^* \cdot {}^{b^"}\underline{I}_{1,Um2} \cdot \exp j\left(\left({}^{b^"}s_{Um2}\omega_{el,Um2} - {}^b s_{Um1}\omega_{el,Um1} \right) t - \left(b^" - b \right) \cdot Z_2\beta_2 \right) \right\},$$

$$(5.4)$$

$$\left(U_{m1} - U_{m2} + \frac{(b^" - b)Z_2}{p} \right) \frac{1}{m} = g, \qquad g = 0, \pm 1, \pm 2, \dots \quad .$$

The rotor circuits are handled in a similar manner:

$$M_{2,I}(t,\beta_2) = \frac{Z_2}{2} \cdot$$

$$\text{Re}\left\{ \sum_{Um1} \sum_{Um2} \sum_v \sum_{v^"} j(v^" - v) \cdot {}^v\underline{\Psi}_{2,Um1} \cdot {}^{v^"}\underline{I}_{2,Um2} \cdot \exp j\left(\left({}^{v^"}s_{Um2}\omega_{el,Um2} + {}^v s_{Um1}\omega_{el,Um1} \right) t + \left(v^" + v \right) \cdot \beta_2 \right) \right\},$$

$$(5.5)$$

$$(v^" + v) = g Z_2, \qquad g = 0, \pm 1, \pm 2, \dots \quad ,$$

$$M_{2,II}(t,\beta_2) = \frac{Z_2}{2} \cdot$$

$$\text{Re}\left\{ \sum_{Um1} \sum_{Um2} \sum_v \sum_{v^"} j(v^" + v) \cdot {}^v\underline{\Psi}_{2,Um1}^* \cdot {}^{v^"}\underline{I}_{2,Um2} \cdot \exp j\left(\left({}^{v^"}s_{Um2}\omega_{el,Um2} - {}^v s_{Um1}\omega_{el,Um1} \right) t + \left(v^" - v \right) \cdot \beta_2 \right) \right\},$$

$$(5.6)$$

$$(v^" - v) = g Z_2, \qquad g = 0, \pm 1, \pm 2, \dots \quad .$$

The flux linkage of the k_{th} stator circuit is the summation over converter implied and machine originated harmonics:

$$\psi_{1,k} = \text{Re}\left\{ \sum_{Um1} \sum_b {}^{b,Um1}\underline{\psi}_{1,k} \right\}, \qquad (5.7)$$

The flux linkage of the k_{th} stator circuit is given by the flux linkage of the first circuit with a phase shift. The term ${}^{b}\underline{\psi}_{1,\text{Um1}} \cdot \sqrt{2}$ is the amplitude of the first circuit.

$$
{}^{b}\overline{\underline{\psi}}_{1,k,\text{Um1}} = \underbrace{\left(\sum_{b'} {}^{b,b'}L_{1,1} \cdot {}^{b'}\underline{I}_{1,\text{Um1}}\sqrt{2} + \sum_{v'} {}^{b,v'}M_{2,1} \cdot {}^{v'}\underline{I}_{2,\text{Um1}}\sqrt{2} \right)}_{{}^{b}\underline{\Psi}_{1,\text{Um1}} \cdot \sqrt{2}}
$$

$$
\cdot \exp j\left[{}^{b}s_{\text{Um1}} \cdot \omega_{\text{Um1}} \cdot t - \left(1 + \frac{bZ_2}{p}\right)(k-1)\frac{2\pi}{m} - bZ_2\beta_2 \right] \quad,
\tag{5.8}
$$

$$
{}^{v}\overline{\underline{\psi}}_{2,a,\text{Um1}} = \underbrace{\left(\sum_{v'} {}^{v,v'}L_{2,2} \cdot {}^{v'}\underline{I}_{2,\text{Um1}}\sqrt{2} + \sum_{b'} {}^{v,b'}M_{1,2} \cdot {}^{b'}\underline{I}_{1,\text{Um1}}\sqrt{2} \right)}_{{}^{v}\underline{\Psi}_{2,\text{Um1}} \cdot \sqrt{2}}
$$

$$
\cdot \exp j\left[{}^{v}s_{\text{Um1}} \cdot \omega_{\text{Um1}} \cdot t - v(a-1)\frac{2\pi}{Z_2} + v\,\beta_2 \right] \quad,
\tag{5.9}
$$

with the inductivities acc. [5.6]:

$$
{}^{b,b'}L_{1,1} = \frac{\mu_0 pq^2 N^2 m l_{\text{Fe}} r}{\pi \delta_g} \cdot {}_{n_2=b'-b}\Lambda_2 \cdot \left[\sum_{v} \frac{{}^{v+b'Z_2}k_w}{v+b'Z_2} \sum_{n_1} \frac{{}^{(v+bZ_2-n_1Z_1)}k_w}{v+bZ_2-n_1Z_1} \right.
$$

$$
\left. {}^{n_1}\Lambda_1 \cdot (-1)^{n_1(mq-\varepsilon-q)} \right] + \left({}^{b}L_{N1} + {}^{b}L_{S1} \right)_{b=b'} \quad,
$$

$$
{}^{v,v'}L_{2,2} = \frac{\mu_0 l_{\text{Fe}} r}{2\delta_g} \cdot \sum_{n_2} \left({}^{n_2}\Lambda_2 \frac{\sin\left(v' - v + n_2 Z_2\right)\frac{\pi}{Z_2}}{v' - v + n_2 Z_2} \right.
$$

$$
\left. \cdot {}^{n_1}\Lambda_1 \cdot (-1)^{n_1(mq-\varepsilon-q)} \right) + \left(2{}^{v}L_{R2} + {}^{v}L_{N2}\,4\sin^2\left(v\frac{\pi}{Z_2}\right) \right)_{v=v'} \quad,
$$

$$
{}^{b,v'}M_{2,1} = \frac{Z_2}{2pm}\frac{\mu_0 pqNml_{\text{Fe}} r}{\pi \delta_g} \cdot \sum_{n_2} \left({}^{n_2}\Lambda_2 \frac{\sin\left(v + b'Z_2 - n_2 Z_2\right)\frac{\pi}{Z_2}}{v + b'Z_2 - n_2 Z_2} \right.
$$

$$
\cdot \sum_{n_1} \left({}^{n_1}\Lambda_1 \frac{{}^{(v+b'Z_2+n_1Z_1)}k_w}{v+b'Z_2+n_1Z_1}(-1)^{n_1(mq-\varepsilon-q)} \right) \quad,
$$

$$
{}^{v,b'}M_{1,2} = \frac{2pm}{Z_2}\,{}^{b,v'}M_{2,1} \quad,
$$

$$
n_1 = 0,\pm1,\pm2,... \quad, \qquad n_2 = 0,\pm1,\pm2,... \quad .
$$

5.2 Calculation of Electromagnetic Torque Oscillations

The flux linkage is in general anti-proportional to the harmonic. In most of the cases mainly the higher stator harmonics, which are coupled to the flux of the base harmonic will contribute to parasitic torque oscillations. If the machine does run at harmonic synchronous speed, the synchronous torque will contribute to the mean value. In principle the analytical theory foresees slot harmonics by permeance functions Λ_1, Λ_2. In case of magnetically closed stator slots the different permeability has to be taken into account for the calculation of the permeance functions. The currents can be calculated according [5.3]. The higher current harmonics are applied to equations 5.3 – 5.6. The result is given in fig. 5.2 for a three-level and multi-level voltage source converter type. Only those torque harmonics are shown, which may influence the design of the shaft. Every torque below the shaft damping is of no interest. All torque oscillations are beneath 10% of the fundamental torque independent upon the applied converter. Only those electrical torque oscillations, which meet one of the shaft line's Eigen-frequencies with the rotor placed at the oscillation loop, require special consideration within the coupling design. Therefore the individual torques need to be applied on the mechanical shaft line system for an appropriate dimensioning of the shaft and coupling [5.7].

The medium voltage converter with low voltage cells show mainly torque oscillations at pulse frequencies. The three-level medium voltage converter has more torque oscillations, fig. 5.2 a). Three main torque oscillation types exist. The third harmonic and parasitic torque are directly related to the fundamental. Torque oscillations due to the pulse frequency are independent from the converter speed. The later occur for the analyzed three-level medium voltage converter in the range of 550 Hz to 800 Hz and 900 Hz to 1100 Hz. In case of the multi-level medium voltage-converter with low voltage cells, oscillations occur at multiples of 5 kHz. They could be adjusted to the individual shaft line situation in extreme cases. The third harmonic and the parasitic torque oscillations can not be influenced, once the machine design and saturation level has been fixed. In some applications it may be possible to change the base frequency of the converter to allow for a reduction of saturation, if the converter is over-dimensioned. Converters with relatively low pulse frequency will cause harmonics, which are directly related to the fundamental frequency. They cause torque oscillations with frequencies, which depend on the motor speed. If the drive is not only operated for some transient moments with these torque components close to the specific Eigen-frequencies, the mechanical stress to the shaft may be too high. Therefore either no torsional Eigen-frequency must occur in the speed range or their amplitude must be lower than the shaft line damping. The influence of these torque oscillations has to be considered for the three-level converter especially for the pulse frequencies between 550 Hz – 1100 Hz. The converter imposed frequencies, which are multiple of the fundamental are below ten percent of the nominal torque. The shaft line should have adequate margins and enough damping to endure these torque oscillations, if the mechanical stress is increased due to resonance of the shaft line. No special inverter influence has to be considered for the medium voltage converter with low voltage cells. The parasitic torque oscillations like in sinusoidal supplied motors are more dangerous. The motor can be regarded as a direct on line operated drive with variable grid frequencies.

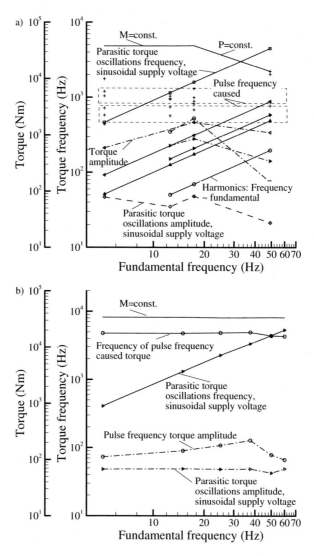

Fig. 5.2 a) Harmonic oscillation for three-level medium voltage converter, b) five-level medium voltage converter with low voltage cells, calculations have been done by adjustments to [5.13]

5.3 Torsional Shaft Oscillations and Design Rules

The converter caused harmonics in the electrical torque will act on the shaft. The mechanical stress depends on the mechanical system behavior. A mathematical model of the shaft line can be determined acc. fig. 5.3.

a) Shaft arrangement

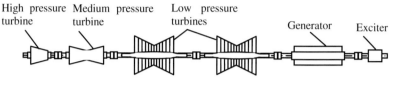

b) Multi-mass-model

c) Multi-mass-model with parallel mass connections

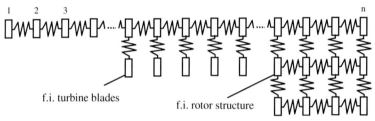

Fig. 5.3 Shaft line model of the mechanical drive configuration of a turbine generator set, acc. [5.7]

Fig. 5.3 gives an overview of a model for a complex shaft arrangement. Industrial motor applications consist often only of four main rotating masses, which are given by the load, the motor and two coupling halves. The inertia moments of the coupling halves are often linked by elements of small stiffness. The coupling is used to trim the shaft Eigen-frequencies.

The electrical torque, which acts on the shaft, is an external torque. Any external torque is either used for the acceleration of the shaft's inertia moments, is converted into damping or transferred to an external torque with negative sign through the individual shaft parts.

The mathematical system of this mechanical model is given in equation 5.12:

$$
\left(J_1 \ \ J_2 \ \cdots \ J_i \ \cdots \ J_n \right) \bullet
\begin{pmatrix} \ddot{\varphi}_1 \\ \ddot{\varphi}_2 \\ \cdot \\ \cdot \\ \ddot{\varphi}_i \\ \cdot \\ \cdot \\ \ddot{\varphi}_n \end{pmatrix}
+
\left(D_{R1} \ \ D_{R2} \ \cdots \ D_{Ri} \ \cdots \ D_{Rn} \right) \bullet
\begin{pmatrix} \dot{\varphi}_1 \\ \dot{\varphi}_2 \\ \cdot \\ \cdot \\ \dot{\varphi}_i \\ \cdot \\ \cdot \\ \dot{\varphi}_n \end{pmatrix}
+
$$

$$
\underbrace{}_{\text{Shaft acceleration}}
\qquad
\underbrace{\phantom{\left(D_{R1} \ \ D_{R2} \ \cdots \ D_{Ri} \right)}}_{\text{Damping proportional speed}}
$$

$$
+
\begin{pmatrix}
\sum\limits_{k=1,n} D_{H1,k} & -D_{H1,2} & \cdots & -D_{H1,k} & \cdots & -D_{H1,n} \\
-D_{H2,1} & \sum\limits_{k=1,n} D_{H2,k} & \cdots & -D_{H2,k} & \cdots & -D_{H2,n} \\
\cdot & \cdot & & \cdot & & \cdot \\
\cdot & \cdot & & \cdot & & \cdot \\
-D_{Hi,1} & -D_{Hi,2} & \cdots & \sum\limits_{k=1,n} D_{Hi,k} & \cdots & -D_{Hi,n} \\
\cdot & \cdot & & \cdot & & \cdot \\
-D_{Hn,1} & -D_{Hn,2} & \cdots & -D_{Hn,i} & \cdots & \sum\limits_{k=1,n} D_{Hn,k}
\end{pmatrix}
\bullet
\begin{pmatrix} \dot{\varphi}_1 \\ \dot{\varphi}_2 \\ \cdot \\ \cdot \\ \dot{\varphi}_i \\ \cdot \\ \dot{\varphi}_n \end{pmatrix}
+
$$

$$
\underbrace{\phantom{\begin{pmatrix} \sum D_{H1,k} & -D_{H1,2} & -D_{H1,n} \end{pmatrix}}}_{\text{Damping proportional speed difference}}
$$

$$
+
\begin{pmatrix}
\sum\limits_{k=1,n} k_{1,k} & -k_{1,2} & \cdots & -k_{1,k} & \cdots & -k_{1,n} \\
-k_{2,1} & \sum\limits_{k=1,n} k_{2,k} & \cdots & -k_{2,k} & \cdots & -k_{2,n} \\
\cdot & \cdot & & \cdot & & \cdot \\
\cdot & \cdot & & \cdot & & \cdot \\
-k_{i,1} & -k_{i,2} & \cdots & \sum\limits_{k=1,n} k_{i,k} & \cdots & -k_{i,n} \\
\cdot & \cdot & & \cdot & & \cdot \\
-k_{n,1} & -k_{n,2} & \cdots & -k_{n,i} & \cdots & \sum\limits_{k=1,n} k_{n,k}
\end{pmatrix}
\bullet
\begin{pmatrix} \varphi_1 \\ \varphi_2 \\ \cdot \\ \cdot \\ \varphi_i \\ \cdot \\ \varphi_n \end{pmatrix}
=
\begin{pmatrix} M_1(t) \\ M_2(t) \\ \cdot \\ \cdot \\ M_i(t) \\ \cdot \\ M_n(t) \end{pmatrix}
$$

$$
\underbrace{\phantom{\begin{pmatrix} \sum k_{1,k} & -k_{1,2} & -k_{1,n} \end{pmatrix}}}_{\text{Torsional stiffness}}
\qquad
\underbrace{\phantom{\begin{pmatrix} M_1(t) \end{pmatrix}}}_{\text{External torques}}
$$

$$
(5.12)
$$

5.3 Torsional Shaft Oscillations and Design Rules

Equation 5.12 is valid for any multi-mass-model with parallel or serial connected masses. Only those stiffness coefficients and hysteresis coefficients are not zero, where a direct connection between two masses exists. In case of a shaft with only series connected masses all coefficients, which are not adjacent to the individual mass, are zero. Parallel connected masses may be considered in case of laminated squirrel cage rotors or turbine blades. The influence of the turbine blades is in the range of 2-3 Hz. The main problem is an appropriate determination of the system parameters. Approaches to calculate inertia moments are given in table 5.2.

Table 5.2 Calculation of inertia-moments

Calculated quantity	Equation	Remark
Inertia moment of a cylinder, acc. [5.8]	$J_x = \dfrac{m\,r^2}{2}$; $J_{z,y} = \dfrac{m\left(3r^2 + h^2\right)}{12}$	with $m = \rho\,\pi\,r^2\,h$
Inertia moment of a hollow cylinder	$J_x = \dfrac{m\left(r_a^2 + r_i^2\right)}{2}$; $J_{z,y} = \dfrac{m\left(3r_a^2 + 3r_i^2 + h^2\right)}{12}$	with $m = \rho\,\pi\left(r_a^2 - r_i^2\right)h$
Inertia moment of a taper with height h	$J_x = \dfrac{3m}{10}\dfrac{\left(r_2^5 - r_1^5\right)}{\left(r_2^3 - r_1^3\right)}$	with $m = \rho\,\pi\cdot h\cdot\left(r_1^2 + r_1 r_2 + r_2^2\right)/3$
Equation of Steiner	$J_x = J_{\bar{x}} + m\left(y_S^2 + z_S^2\right)$	with y_S, z_S distance of x axis to center of gravity

Beside the inertia moment the stiffness coefficient has to be determined. Whereas this coefficient can be relatively simple calculated for massive rotors, a tapered shaft end or a feather key is more complex. Basic rules and sample values are given in table 5.3.

Last but not least the damping has to be determined. The damping is split in two kinds of damping. Friction losses are proportional to the speed. They are caused by ventilation losses as well as by surface losses and bearing losses. A challenge is the appropriate determination of the shaft inherent damping. The hysteresis damping coefficients are material inherent or given by laminations, which rub on each other. An overview of typical coefficients is given in table 5.4. The inherent damping is always attached to special Eigen-formen. In [5.7] a determination approach for the modal damping coefficient is given. A main challenge is to derive out of measurements from the complete system the individual relationship between two adjacent masses.

80   5 Converter Caused Torque Oscillations

Table 5.3 Determination of stiffness coefficients

Calculated quantity	Equation	Remark
Stiffness	$$k_t = \frac{G I_P}{L}$$	with L, length of shaft element; Steel constant: $G=80000$ MN/m^2
Stiffness for shafts with feather key	$$k_t = \frac{G I_P(d')}{\kappa \cdot L};\ d'=d-h;\ \kappa=1;\ h/d' \le 0.05$$ $$\kappa = 0.96 + 0.8\,h/d';\ h/d' \ge 0.05$$	with h height of feather key, d shaft diameter
Polar geometrical moment of inertia for cylindrical designs	$$I_P = \frac{\pi\, d^4}{32}$$	with d cylinder diameter
Polar geometrical moment of inertia for tapered designs	$$I_P = \frac{3\pi\,(d_2 - d_1)}{32}\,\frac{d_2^3 d_1^3}{\left(d_2^3 - d_1^3\right)}$$	with d_1, d_2 taper diameters
Modification of L by $\pm \Delta l$ at transitions between two cylinders with different diameters d_k, d_{k+1}, acc. [5.9, 5.10]	$$\Delta l / d_k = 0.155;\ 0 \le d_k / d_{k+1} \le 0.3$$ $$\Delta l / d_k = -0.155 \frac{d_k}{d_{k+1}} + 0.1895;$$ $$0.3 \le d_k / d_{k+1} \le 0.5$$ $$\Delta l / d_k = 0.264 \left(1 - \frac{d_k}{d_{k+1}}\right);$$ $$0.5 \le d_k / d_{k+1} \le 1$$	the additional length will be added for the smaller diameter and subtracted for the larger diameter, no transition radius

The inherent damping depends upon the Eigen-form. Only those shaft parts contribute, which are exposed to maximal stress values or potential energies. The damping acc. table 5.4 defines a modal damping related to one special Eigen-form i. In order to determine the modal inertia moment and modal frequency equation 5.12 has to be translated in its modal form. This is done in the following steps according equations 5.13-5.14:

$$(J)\left(\ddot{\varphi}\right) + (k)(\varphi) = 0 \tag{5.13}$$

$$\underbrace{(X_T)(J)(X)}_{J_m} \cdot \left(\ddot{q}\right) + \underbrace{(X_T)(D)(X)}_{D_m} \cdot \left(\dot{q}\right) + \underbrace{(X_T)(k)(X)}_{k_m}(q) = \underbrace{(X_T)(M)}_{M_m}, \tag{5.14}$$

with (X) Matrix of Eigen-vectors,

5.3 Torsional Shaft Oscillations and Design Rules

Table 5.4 Determination of damping coefficients [5.7, 5.9, 5.11]

Calculated quantity	Equation	Remark
Logarithmic decrement	$\Lambda = \ln \dfrac{\varphi_n}{\varphi_{n+1}} = \delta \cdot T_d$ $T_d = \dfrac{2\pi}{\omega_d} \approx \dfrac{2\pi}{\omega_0}$	decay constant δ, Eigen-frequency with/without damping: ω_d / ω_0
Damping D	$D = \delta \cdot 2 \cdot J_i = \dfrac{\omega_i}{\pi} \Lambda \cdot J_i$	
Typical damping coefficients for special applications	Compressors, fans: $D_a = 19.1 \cdot \dfrac{M_n}{n}$ Ship propellers $D_a = 38.2 \cdot \dfrac{M_n}{n}$	M_n in Nm, n in rpm
Inherent shaft Damping, logarithmic decrement	Shaft: $\Lambda = 0.03...0.06$ Gear-step <100kW: $\Lambda = 0.13$ Gear-step 100kW-1MW: $\Lambda = 0.25$ Gear-step >1MW: $\Lambda = 0.38$ Elastic coupling: $\Lambda = 0.13...1.26$	

$$(X_T)(D)(X) = \begin{pmatrix} D_1 & \cdots & 0 & \cdots & 0. \\ \cdot & \cdot & \cdot & \cdot & \cdot \\ \cdot & \cdot & \cdot & \cdot & \cdot \\ \cdot & \cdot & \cdot & \cdot & \cdot \\ 0 & \cdots & D_i & \cdots & 0 \\ \cdot & \cdot & \cdot & \cdot & \cdot \\ \cdot & \cdot & \cdot & \cdot & \cdot \\ \cdot & \cdot & \cdot & \cdot & \cdot \\ 0 & \cdots & 0 & \cdots & D_n \end{pmatrix}, \quad (q) = (X^{-1})(\varphi), \quad (\varphi) = (X)(q).$$

n-1 Eigen-frequencies do exist for rotating shafts. The missing Eigen-vector can be derived for $\omega_1 = 0$. The Eigen-vector is given by: $(x_1) = (1 \quad ... \quad 1 \quad ... \quad 1)$. The damping coefficient D_i is only clearly determinable, if the shaft is operated in the individual Eigen-frequency.

A torsional shaft analysis can be calculated after the necessary shaft parameters have been determined. An example for the frequency dependent torsional oscillation amplitude, which is related to the amplitude at zero frequency, is given for a four mass system in fig. 5.4.

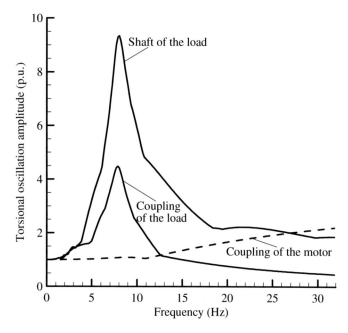

Fig. 5.4 Torsional oscillation levels as function of frequency including the first Eigen-form of a four mass system with load, motor and two coupling halves

The modal damping limits directly the maximal amplitudes in the individual Eigen-frequencies. The inertia moment is the summation of those masses, which are involved on the oscillation mode. The system has modes at 8.0 Hz, 180 Hz, and 256 Hz. The last two modes are modes of the coupling halves. Only frequencies up to 32 Hz are shown in fig. 5.4. Only the resonance peak of the mode motor versus load is shown. Strongest oscillation values against the motor can be seen at the load position. One coupling oscillates with the load, the other with the motor.

The mechanical equation system has been solved in modal form and the coordinates q are transformed for every time step afterwards in order to include the modal damping. Fig. 5.5 shows a comparison between the modal coordinates and the actual shaft angles.

The coordinate, which is related to the individual Eigen-mode, oscillates in modal coordinates 5.5. a). The actual shaft movement is related to all four masses, fig. 5.5 b).

5.3 Torsional Shaft Oscillations and Design Rules

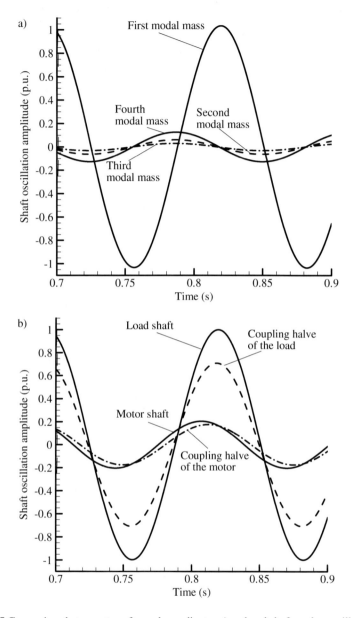

Fig. 5.5 Comparison between transformed coordinates a) and real shaft angles oscillations b)

Based upon the determination of the individual mass rotation angle differences, individual material stresses can be determined acc. equations 5.15-5.18.

$$M_t = k_{i,i+1} \cdot \left(\varphi_i - \varphi_{i+1} \right), \tag{5.15}$$

$$M_t = \int_{d_i/2}^{d_a/2} \tau(r) \cdot r \, dA = \left(G \cdot \frac{\phi}{l} \right) \cdot I_P, \tag{5.16}$$

$$\tau(r) = G \frac{\phi}{l} r = \frac{M_t}{I_P} \cdot r, \tag{5.17}$$

$$\tau_{max} = \frac{M_t}{I_P} \cdot r_{max} = \frac{M_t}{I_P} \cdot \frac{d_a}{2}, \tag{5.18}$$

with a transposition per length ϕ/l.

The maximum material stress according equation 5.18 does not take into account actual dimension changes, feather keys nor material influences. Design and manufacture specific margins have to be considered for these influences. The maximum amplitude of the resulting equivalent stress has to be below the infinite load cycles for all operation frequencies. If the equivalent stresses are too high, special measures have to be considered. The torsional load for the coupling and shaft can be handled in general by an appropriate material and dimensional choice. Special kinds of couplings like oil shrink fits could be used for instance, when a standard feather key would be at its end.

Chapter 6
Noise Based on Electromagnetic Sources in Case of Converter Operation

6.1 General Overview of Noise Calculation

The pre-calculation of noise is one of the most challenging thematic within the design of electrical machines. Noise pressure levels have often to be guaranteed to be f.i. lower than 82 dB or 85 dB on load. Up to today no real calculation method exists to calculate the overall noise with an accuracy of 1-3 dB. A principle approach to noise calculation is given in fig. 6.1.

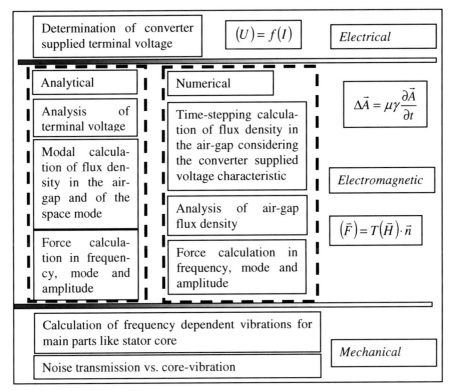

Fig. 6.1 Calculation approach for the determination of the noise level in converter supplied drives

Noise sources can be of different kind. The ventilation schema and mechanical vibrations due to shaft vibrations are noise sources beside electromagnetic sources. They are often the dominant sources in sinusoidal supplied air cooled 2- and 4-pole machines, but they are independent upon the converter supply. Therefore ventilation - and mechanically caused noise will not be handled in this work. The noise transmission within the mechanical machine structure and through the air-gap will be considered.

The calculation schema can be structured in three main parts. Within a first step the terminal voltage has to be determined. The terminal voltage is then applied to an analytical [6.1] or numerical machine model in order to calculate electromagnetic forces. Last but not least the forces are applied to a mechanical model of the stator core. The vibrations of the core are transmitted towards machine surfaces and the air, where the pressure variation will be called noise. A complete numerical calculation procedure of all three parts has been performed in [6.2]. Often a focus to main effects is important to understand the phenomena. The following focus will be on those force harmonics, which have been the main cause of noise in different investigations. A more complete set of harmonics may be found in literature [6.1, 6.4-6.6].

6.2 Determination of the Electromagnetic Force Modes and Amplitudes

The converter supplied terminal voltage depends strongly on the type of converter. It may be even necessary to integrate the power electronic components in the machine model in order to get the correct stator currents in case of current source converters [6.7, 6.8]. Unfortunately the exact converter pulse muster of active power electronic elements is one of the main core competences within the converter design of manufacturers and hardly published. Even though only rare opportunities exist to calculate the correct terminal voltage by circuit models in principle it is possible if main capacitive effects are taken into account. Main focus is given on the correct simulation of the converter. In case of current source converters the commutation of the power electronic depends strongly on the leakage reactance of the motor. A more detailed model taking saturation into account improves the results especially if different stator current levels due to load variation are investigated. In case of voltage source converters it is sufficient to model the machine relatively roughly with circuit elements. The influence of a detailed machine model on the terminal voltage is relatively low.

Even more reliable results can be obtained by measurements of the terminal voltage in the plant, where all capacitive effects are considered properly. It is important to measure the voltage between the terminal and the neutral point in case of star connection. The neutral point should be installed in a separate terminal box in this case. Measurements within test facilities may be rechecked with simulations where the correct cable length and capacity can be considered.

Based on the terminal voltage the machine currents and the acting forces on the stator and rotor core or teeth can be calculated. The calculation is mainly based on

6.2 Determination of the Electromagnetic Force Modes and Amplitudes

the flux density distribution within the machine's air-gap. Two principle methods for the determination of the flux distribution can be considered. In literature the calculation is mostly based on analytical methods, which model the machine by the superposition of harmonics [6.5, 6.6]. The principles of this method have been demonstrated in line with the torque oscillation calculation. The main difference is given in the harmonics, which need to be considered. Whereas only those harmonic components, which have the same number of pole pairs contribute to a torque or torque oscillation on the shaft, harmonic components, which are different in pole pairs contribute to noise as well.

In case of the noise calculation within this chapter a method based on numerical time-stepping finite difference calculation will be demonstrated. The theoretical basics have been elaborated in chapter two. The method considers the rotor rotation as well as saturation effects and the transient voltage form as a whole. A typical distribution of the radial flux density is given in fig. 6.2.

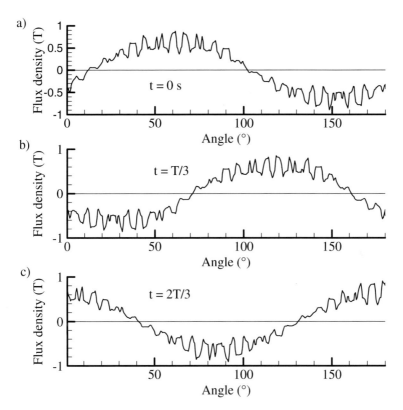

Fig. 6.2 Radial flux density distribution within the air-gap of a 4-pole, converter driven machine for different time steps in the stator coordinate system

The flux density distribution on its own does not allow any estimation on its effects towards noise, but on the possible modal forms. Mainly the change in time will cause vibrations, but the modal forms determine the transmission function of the mechanical structure. Within steady state operation the flux density distribution in the air-gap is periodic in time and space. The flux density is given by the following equation 6.1:

$$B(x,t) = \sum_{v}\sum_{\mu} \hat{B}_{-,v,\mu,\cos} \cdot \cos(v \cdot \varphi - \mu \cdot \omega_{el} \cdot t) + \sum_{v}\sum_{\mu} \hat{B}_{+,v,\mu,\cos} \cdot \cos(v \cdot \varphi + \mu \cdot \omega_{el} \cdot t)$$
$$+ \sum_{v}\sum_{\mu} \hat{B}_{-,v,\mu,\sin} \cdot \sin(v \cdot \varphi - \mu \cdot \omega_{el} \cdot t) + \sum_{v}\sum_{\mu} \hat{B}_{+,v,\mu,\sin} \cdot \sin(v \cdot \varphi + \mu \cdot \omega_{el} \cdot t) \quad ,$$

$$(6.1)$$

with fundamental frequency ω_{el}, harmonic-amplitude $\hat{B}_{v,\mu}$, $v = 0,1,2,3,4...$, $\mu = 1,2,3,4...$.

The equation can be transformed by using equation 6.2:

$$\cos(v \cdot \varphi - \mu \cdot \omega_{el} \cdot t) = \cos(v \cdot \varphi) \cdot \cos(\mu \cdot \omega_{el} \cdot t) + \sin(v \cdot \varphi) \cdot \sin(\mu \cdot \omega_{el} \cdot t)$$
$$\cos(v \cdot \varphi + \mu \cdot \omega_{el} \cdot t) = \cos(v \cdot \varphi) \cdot \cos(\mu \cdot \omega_{el} \cdot t) - \sin(v \cdot \varphi) \cdot \sin(\mu \cdot \omega_{el} \cdot t)$$
$$\sin(v \cdot \varphi - \mu \cdot \omega_{el} \cdot t) = \sin(v \cdot \varphi) \cdot \cos(\mu \cdot \omega_{el} \cdot t) - \cos(v \cdot \varphi) \cdot \sin(\mu \cdot \omega_{el} \cdot t)$$
$$\sin(v \cdot \varphi + \mu \cdot \omega_{el} \cdot t) = \sin(v \cdot \varphi) \cdot \cos(\mu \cdot \omega_{el} \cdot t) + \cos(v \cdot \varphi) \cdot \sin(\mu \cdot \omega_{el} \cdot t) \quad .$$

$$(6.2)$$

Equation 6.1 is split in time dependent and space dependent terms:

$$B(x,t) = \sum_{v}\sum_{\mu} \hat{B}_{-,v,\mu,\cos} \cdot (\cos(v \cdot \varphi) \cdot \cos(\mu \cdot \omega_{el} \cdot t) + \sin(v \cdot \varphi) \cdot \sin(\mu \cdot \omega_{el} \cdot t))$$
$$+ \sum_{v}\sum_{\mu} \hat{B}_{+,v,\mu,\cos} \cdot (\cos(v \cdot \varphi) \cdot \cos(\mu \cdot \omega_{el} \cdot t) - \sin(v \cdot \varphi) \cdot \sin(\mu \cdot \omega_{el} \cdot t))$$
$$+ \sum_{v}\sum_{\mu} \hat{B}_{-,v,\mu,\sin} \cdot (\sin(v \cdot \varphi) \cdot \cos(\mu \cdot \omega_{el} \cdot t) - \cos(v \cdot \varphi) \cdot \sin(\mu \cdot \omega_{el} \cdot t))$$
$$+ \sum_{v}\sum_{\mu} \hat{B}_{+,v,\mu,\sin} \cdot (\sin(v \cdot \varphi) \cdot \cos(\mu \cdot \omega_{el} \cdot t) + \cos(v \cdot \varphi) \cdot \sin(\mu \cdot \omega_{el} \cdot t)) \quad ,$$

$$(6.3)$$

with fundamental frequency ω_{el}, harmonic-amplitude $\hat{B}_{v,\mu}$, $v = 0,1,2,3,4...$, $\mu = 1,2,3,4...$.

6.2 Determination of the Electromagnetic Force Modes and Amplitudes

$$B(x,t) = \sum_{\nu}\sum_{\mu}\cos(\nu\cdot\varphi)\cdot$$

$$\cdot\left(\left(\hat{B}_{-,\nu,\mu,\cos} + \hat{B}_{+,\nu,\mu,\cos}\right)\cdot\cos(\mu\cdot\omega_{el}\cdot t) + \left(\hat{B}_{+,\nu,\mu,\sin} - \hat{B}_{-,\nu,\mu,\sin}\right)\sin(\mu\cdot\omega_{el}\cdot t)\right)$$

$$+ \sum_{\nu}\sum_{\mu}\sin(\nu\cdot\varphi)\cdot$$

$$\cdot\left(\left(\hat{B}_{-,\nu,\mu,\cos} - \hat{B}_{+,\nu,\mu,\cos}\right)\cdot\sin(\mu\cdot\omega_{el}\cdot t) + \left(\hat{B}_{+,\nu,\mu,\sin} + \hat{B}_{-,\nu,\mu,\sin}\right)\cos(\mu\cdot\omega_{el}\cdot t)\right) \quad ,$$

$$(6.4)$$

with fundamental frequency ω_{el}, harmonic-amplitude $\hat{B}_{\nu,\mu}$, $\nu = 0,1,2,3,4...$, $\mu = 1,2,3,4...$.

On the basic form of equation 6.4 it is possible to Fourier-analyse the flux distribution in space for any time step. The harmonics in time can be Fourier-analysed based on this result in order to get the harmonic amplitudes:

$$\left(\hat{B}_{-,\nu,\mu,\cos} + \hat{B}_{+,\nu,\mu,\cos}\right), \ \left(\hat{B}_{+,\nu,\mu,\sin} - \hat{B}_{-,\nu,\mu,\sin}\right), \ \left(\hat{B}_{-,\nu,\mu,\cos} - \hat{B}_{+,\nu,\mu,\cos}\right),$$

$$\left(\hat{B}_{+,\nu,\mu,\sin} + \hat{B}_{-,\nu,\mu,\sin}\right).$$

Travelling waves in the positive direction are given by:

$$\left(\hat{B}_{\nu,\mu,pos} = \sqrt{\hat{B}^2_{-,\nu,\mu,\cos} + \hat{B}^2_{-,\nu,\mu,\sin}}\right) .$$

Travelling waves in the negative direction are given by:

$$\left(\hat{B}_{\nu,\mu,neg} = \sqrt{\hat{B}^2_{+,\nu,\mu,\cos} + \hat{B}^2_{+,\nu,\mu,\sin}}\right).$$

The analysis of the machine with the flux density distribution acc. fig. 6.2 is given in fig. 6.3. Additional the time characteristic is given for the fundamental space distribution, the saturation mode and the rotor slot harmonics in fig. 6.4.

The transients of the calculation with a time-stepping method are not completely decayed and can be seen slightly in the characteristics. Reasons for main harmonics within the air-gap flux density are given in table 6.1 for forces acting on the stator teeth. Real machines will have at least static eccentricities, which will influence occurring modes and harmonics. This has been taken into account.

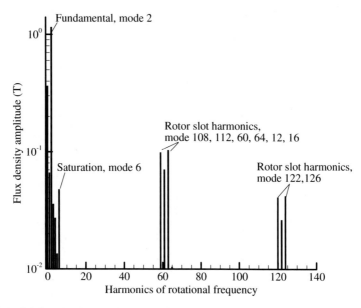

Fig. 6.3 Radial flux density amplitudes of frequency harmonics in the air-gap from a 4-pole asynchronous motor, operation with a three-level converter, related to the stator coordinate system

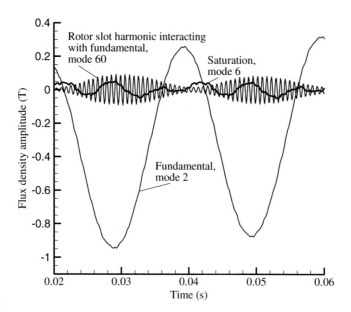

Fig. 6.4 Radial flux density amplitudes versus time for the fundamental space harmonic, rotor slot harmonics and saturation harmonic

6.2 Determination of the Electromagnetic Force Modes and Amplitudes

Table 6.1 Dominant harmonics in the air-gap flux density

Mode	Frequency Stator	Frequency Rotor	Amplitude (%)	Reason
P	f_{el}	0	100	Stator field fundamental
$P\pm1$	f_{el}	f_{el}/p	2	Eccentricity and fundamental
$3p$	$3\,f_{el}$	$2\cdot f_{el}$	5	
$5p$	f_{el}	$6\cdot f_{el}$	1.2	Stator field harmonics
$7p$	f_{el}	$6\cdot f_{el}$	0.75	
$11p$	f_{el}	$12\cdot f_{el}$	0.85	
$13p$	f_{el}	$12\cdot f_{el}$	0.88	
$v\cdot p$	f_{el}	$2ma\cdot f_{el}$		Stator field harmonics, $v = p\cdot(2ma+1)$, $a = 0, \pm1, \pm2, ..., \pm i$, with number of stator phases m
$Z_1\text{-}p$	f_{el}	$Z_1/p\cdot f_{el}$	12	Stator slot harmonics
$Z_1\text{+}p$	f_{el}	$Z_1/p\cdot f_{el}$	15	
$Z_1\text{-}p\text{-}1$	f_{el}	$(Z_1/p-1/p)\cdot f_{el}$	0.5	Stator slot harmonics and eccentricity
$Z_1\text{+}p\text{-}1$	f_{el}	$(Z_1/p-1/p)\cdot f_{el}$	0.5	
$Z_1\text{+}p\text{+}1$	f_{el}	$(Z_1/p-1/p)\cdot f_{el}$	0.5	
$Z_1\text{-}p\text{+}1$	f_{el}	$(Z_1/p-1/p)\cdot f_{el}$	0.5	
$Z_2\text{+}p$	$(Z_2/p+1)\cdot f_{el}$	0	9	Rotor slots, fundamental
$Z_2\text{-}p$	$(Z_2/p-1)\cdot f_{el}$	0	8	Rotor slots, fundamental
$Z_1\text{-}Z_2\text{+}p$	$(Z_2/p-1)\cdot f_{el}$	$(Z_1/p+1)\cdot f_{el}$	1.4	Stator and rotor slots modulate the fundamental
$Z_1\text{-}Z_2\text{-}p$	$(Z_2/p+1)\cdot f_{el}$	$(Z_1/p-1)\cdot f_{el}$	0.028	
$2\cdot Z_1\text{-}Z_2\text{+}p$	$(Z_2/p-1)\cdot f_{el}$	$Z_1/p\cdot f_{el}$	0.0094	Second stator, rotor slot harmonic

The magnetic force density distribution which acts on the teeth can be derived in the same way. The result is given in equation 6.5:

$$f(x,t) = \sum_{v}\sum_{\mu} \cos(v \cdot \varphi) \cdot$$

$$\cdot \left(\left(\hat{f}_{-,v,\mu,\cos} + \hat{f}_{+,v,\mu,\cos} \right) \cdot \cos(\mu \cdot \omega_{el} \cdot t) + \left(\hat{f}_{+,v,\mu,\sin} - \hat{f}_{-,v,\mu,\sin} \right) \sin(\mu \cdot \omega_{el} \cdot t) \right)$$

$$+ \sum_{v}\sum_{\mu} \sin(v \cdot \varphi)$$

$$\cdot \left(\left(\hat{f}_{-,v,\mu,\cos} - \hat{f}_{+,v,\mu,\cos} \right) \cdot \sin(\mu \cdot \omega_{el} \cdot t) + \left(\hat{f}_{+,v,\mu,\sin} + \hat{f}_{-,v,\mu,\sin} \right) \cos(\mu \cdot \omega_{el} \cdot t) \right),$$

$$(6.5)$$

with fundamental frequency ω_{el}, harmonic-amplitude $\hat{f}_{v,\mu}$, $v = 0,1,2,3,4...$, $\mu = 1,2,3,4...$.

The magnetic force density is calculated based on the magnetic stress tensor T_m:

$$T_m = \begin{bmatrix} \mu H_x^2 - \frac{1}{2}\mu H^2 & \mu H_x H_y & \mu H_x H_z \\ \mu H_y H_x & \mu H_y^2 - \frac{1}{2}\mu H^2 & \mu H_y H_z \\ \mu H_z H_x & \mu H_z H_y & \mu H_z^2 - \frac{1}{2}\mu H^2 \end{bmatrix}. \qquad (6.6)$$

The force density is given by the product of the magnetic stress tensor with the surface normal vector $\vec{n}$. In case of a two-dimensional field distribution is the axial component of the field strength zero. The surface normal can be chosen to be

$\vec{n} = \begin{pmatrix} 1 \\ 0 \\ 0 \end{pmatrix}$ with an appropriate coordinate system of the air-gap field:

$$T_m \cdot \vec{n} = \begin{bmatrix} \mu H_x^2 - \frac{1}{2}\mu H^2 & \mu H_x H_y & 0 \\ \mu H_y H_x & \mu H_y^2 - \frac{1}{2}\mu H^2 & 0 \\ 0 & 0 & -\frac{1}{2}\mu H^2 \end{bmatrix} \cdot \begin{pmatrix} 1 \\ 0 \\ 0 \end{pmatrix} = \begin{pmatrix} \mu H_x^2 - \frac{1}{2}\mu H^2 \\ \mu H_y H_x \\ 0 \end{pmatrix}.$$

$$(6.7)$$

In case of the surface of iron with perpendicular field distribution only the radial field component contributes towards the acting forces. Indeed the component $\mu H_y H_x$ contributes mainly to the machine torque. Both components are shown in fig. 6.5 and 6.6.

6.2 Determination of the Electromagnetic Force Modes and Amplitudes

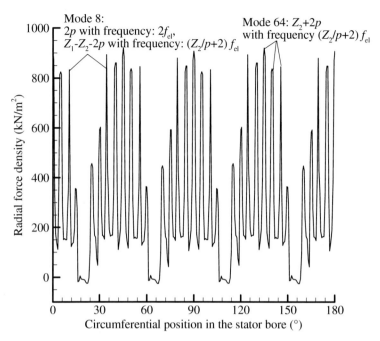

Fig. 6.5 Radial force density of an 8-pole machine, 56 rotor and 72 stator slots

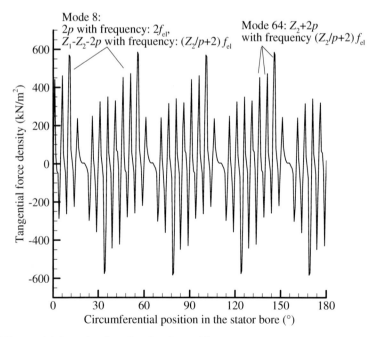

Fig. 6.6 Tangential force density of an 8-pole machine

The forces are caused by two flux density components. These components are travelling waves with a function description according equation 6.8:

$$\cos(\nu_1 \cdot \varphi - \mu_1 \cdot \omega_{el} \cdot t) \cdot \cos(\nu_2 \cdot \varphi - \mu_2 \cdot \omega_{el} \cdot t)$$
$$= \left(\underbrace{\cos((\nu_1 + \nu_2) \cdot \varphi - (\mu_1 + \mu_2) \cdot \omega_{el} \cdot t)}_{Part1} + \underbrace{\cos((\nu_1 - \nu_2) \cdot \varphi - (\mu_1 - \mu_2) \cdot \omega_{el} \cdot t)}_{Part2} \right) / 2$$
(6.8)

Part 1 is the more important part for noise, only part 2 is relevant for torque oscillations, if both space harmonics are of the same value. Obviously a fourth time harmonic can exist in the noise spectrum, but the same flux density distribution would cause a sixth time harmonic in the torque oscillations.

Beside the relative large modes of the slot harmonics a mode eight is caused by fundamental space harmonic and by the stator-rotor slot difference harmonics. Odd modes can be caused by machine asymmetries. One asymmetry, which occurs always, is the eccentricity of the rotor in the stator bore. Due to this eccentricity additional modes are caused. Fig. 6.7 shows an example for the force modulation.

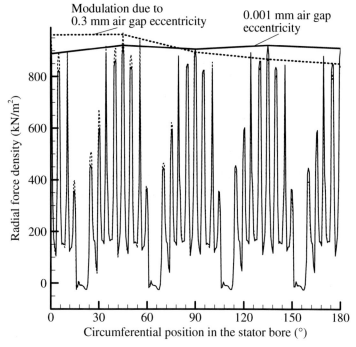

Fig. 6.7 Radial force density of the positive travelling wave for a three-level converter on the rotor surface, rotating coordinate system

Asynchronous machines in the size between 500 kW to about 15 MW have air-gaps of 1.5 mm to 5 mm. Larger machines for higher speeds do have larger air-gaps, machines with six and more poles have smaller air-gaps. Normal asynchronous machines should not have eccentricities of much more than 0.3mm. The eccentricity according fig. 6.7 causes a mode 7 and a mode 9. Mode 7 will have a frequency of twice the stator voltage supply frequency and the rotor slot modulated frequency. The amplitude is relative small in the range of 5000 N/m^2.

6.3 Eigen-Modes and Mechanical Calculation Methods

The radial forces cause vibrations of the stator or rotor core. The vibration will be a mixture of mechanical Eigen-modes. An overview of mechanical Eigen-modes is given in fig. 6.8. The stator gets a certain form, which defines the mode acc. literature [6.3, 6.6]. Often only free symmetric stators are shown. Indeed the mechanical design becomes asymmetric due to the connection of the motor to the foundation. One typical mode is split f.i in three. Only one of the multiple modes is shown in fig. 6.8 for each of the typical forms.

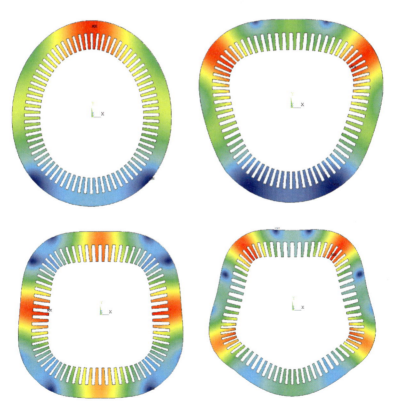

Fig. 6.8 Vibration modes 2, 3, 4 and 5, taking the connection to the foundation into account

The calculation of the individual mechanical vibration amplitudes consists on the one hand of the determination of the Eigen-frequencies for the modes r of the mechanical structure and on the other hand on the determination of their amplitudes.

The structure mechanical calculation has been described in literature [6.1, 6.3, 6.9]. Different approaches can be found in order to determine Eigen-frequencies and modes. Typical equations for Eigen-frequencies and amplitudes are well known for cylindrical shells [6.1]:

$$f_{b,r} = \frac{f_0}{2\sqrt{3}} \frac{h_y}{R_m} \frac{\left(r\left(r^2-1\right)\right)}{\sqrt{r^2+1}}, \tag{6.9}$$

$$\hat{Y}_r = 12 \frac{R_\delta}{E} \left(\frac{R_m}{h_y}\right)^3 \frac{\hat{T}_{m,y,r}}{\left(r^2-1\right)^2} \frac{1}{\left|1-\eta^2\right|}, \tag{6.10}$$

with modes r, stator yoke height h_y, radius to the middle of the yoke R_m, amplitude of the magnetic stress in radial direction $\hat{T}_{m,y,r}$, the frequency of the mode zero:

$$f_0 = \frac{\sqrt{E/\rho}}{2\pi \cdot R_m \sqrt{\Delta}},$$ the amplitude of the mode zero: $\hat{Y}_0 = \frac{R_\delta}{E} \frac{R_m}{h_y} \hat{T}_{y,0} \frac{1}{\left|1-\eta^2\right|},$ the

relationship between the frequency of the excitation force and the closest Eigen-frequency $\eta = f_{force}/f_{b,r}$, the influence of the stator teeth as additional mass is given by Δ, which is the mass of stator yoke and teeth divided by the mass of stator yoke.

The vibration level is dominated by the damping of the mechanical structure at the Eigen-frequency. Values for the damping are given in [6.1]. The relatively simple equations allow a reasonable agreement with measurements [6.1], if some assumptions are fulfilled.

- the stator winding is of randomly wound wire design, which does not increase the stator stiffness,
- teeth Eigen-modes are far away from the dominant vibrations,
- the mechanical vibration mode is excited by a force of the same mode,
- the noise is dominated by forces, which act directly on the stator.

Teeth Eigen-modes have a strong influence on Eigen-frequencies of higher order. A comparison of the relative simple equation 6.9 with a numerical core model without stator winding has shown deviations of less than fifteen percent up to the fourth mode. Higher modes are in most of industrial applications not really responsible for too high noise levels. The amplitude reduces nearly by the power of four with the mode. The influence of the stator teeth on an analytical solution can be found in [6.9]. The influence of the stator-winding is much more important. Electrical machines with form wound coils do not fulfil the assumption, that the

6.3 Eigen-Modes and Mechanical Calculation Methods   97

winding does not increase the core stiffness. The vacuum pressure impregnated coils and the slot wedges will support the teeth in tangential and radial direction. Table 6.2 gives an example, where the application of simple equations acc. 6.9 is no longer valid:

Table 6.2 Comparison of different calculation methods with measurements

Mode	Analytically calculated frequency, acc. [6.1]	Numerical calculated frequency taking the winding stiffness into account	Dominant measured frequency with slip accuracy
0	1760 Hz	1730 Hz	-
2	211 Hz	305 Hz	-
3	596 Hz	808 Hz	900 Hz
4	1142 Hz	1447 Hz	1500 Hz
5	1847 Hz	2184 Hz	-

It is obvious that the analytical solution deviates too much from the measurement. The main benefits of the analytical equations are the basic principles, which can be derived out of the simple equations:

- Influence of the inner radius
- Influence of the joke
- Influence of the mode

The machine structure including the winding becomes too complex for analytical calculation methods. A proper calculation of modes and amplitudes can only be realized by numerical methods. It is essential to consider the insulation stiffness correctly. The stiffness of the insulation of form wound machines may influence the mode frequencies by up to 30%. The influence may be less in randomly wound smaller machines. Analytical methods may give good results even for the third mode, [6.1].

Even though that the measured frequency has been correlated to the third mechanical mode with 808 Hz some questions do still exist. The measured machine is an 8-pole asynchronous machine with 72 stator slots and 56 rotor slots. Neither a third mode nor 900 Hz can be found at 50 Hz operation in the magnetic force. The 72 stator slots will create forces which act with 900 Hz on the rotor, but not on the stator. Literature about noise in electrical machines deals only with modes and frequencies of forces acting on the stator. Indeed the rotor-surface is coupled with the stator over the air-gap. The air of the air-gap is not capable to vanish in axial directions at frequencies of 500 Hz up to 10 kHz. Indeed the air is reacting like air in a piston air cylinder, [6.10, 6.11]. The phenomenon has two effects. On the one hand the rotor surface forces are coupled to the stator by the stiffness of the air-gap. On the other hand the machine has to be considered as a coupled structure between rotor and stator, see fig. 6.9.

The stiffness can be determined acc. equation 6.11:

$$C' = \frac{\rho c^2}{\delta},\qquad(6.11)$$

with C': Stiffness per surface, c: sonic speed, δ: air-gap; ρ density of air.

The assumption that the air-gap is a mere stiffness element is valid as long as the following relationship 6.12 is valid:

$$\frac{\omega}{c}\cdot\delta < 1,\qquad(6.12)$$

with vibration angular frequency ω.

Larger frequencies or air-gaps would cause reflections. The stiffness of the air-gap has been included in the numerical model acc. fig.6.9.

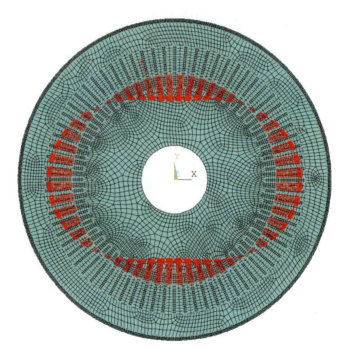

Fig. 6.9 Numerical model where magnetic forces act on the rotor

The described measurement is a very special case. Forces, which act on the rotor, can only cause noise components, if the rotor and stator Eigen-frequencies of some modes are in the same range. If the Eigen-frequencies are different, the displacement amplitudes at the rotor surfaces are very small for an excitation with the stator Eigen-frequency. The forces, which act on the stator bore depend directly on the rotor surface displacement and will be quite small as well.

6.3 Eigen-Modes and Mechanical Calculation Methods

Force modes and mechanical Eigen-modes have to be differentiated strongly. Mechanical modes of cylinders will be deformed in the circumferential direction with a pure sinusoidal characteristic. Motors, which are clamped to a foundation for instance, will become asymmetric from a mechanical point of view. The individual vibration modes will no longer be described by one sinusoidal distribution but by a summation of space harmonics. If any of these space harmonics would be excited with the Eigen-frequency of the mode, the mechanical mode will occur. The excitation of a certain Eigen-mode can be achieved by any magnetic force mode which is included in the mechanical vibration form and has an appropriate frequency. Even though the dominant mode within a mechanical Eigen-form may be described by a mode three, it can be excited by an electromagnetic force of a mode 8. This factor will be investigated in more detail. Fig. 6.10 shows the transfer functions from rotating magnetic forces of different modes to the stator frame of a horizontal mounted motor, which is fixed on a foundation.

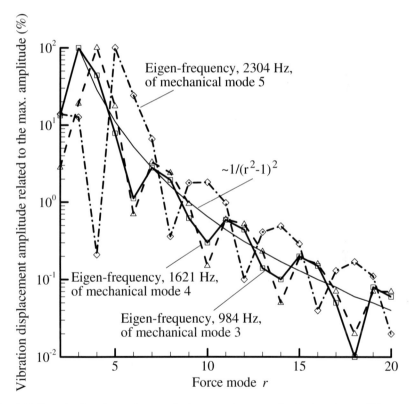

Fig. 6.10 Transfer functions of magnetic force modes to mechanical modes at different Eigen-frequencies for a horizontal mounted motor

Beside the above described vibration of the stator core other vibrations occur, which are often not in the focus, if electromagnetic noise is discussed. The typical transformer grumble is mainly caused by vibrations of the lamination. Again the electromagnetic stress tensor can be used for this force calculation. It is assumed, that the flux is led by the lamination.

The stress tensor will describe the forces on the surface:

$$T_m \cdot \vec{n} = \begin{bmatrix} \mu H_x^2 - \dfrac{1}{2}\mu H^2 & \mu H_x H_y & 0 \\ \mu H_y H_x & \mu H_y^2 - \dfrac{1}{2}\mu H^2 & 0 \\ 0 & 0 & -\dfrac{1}{2}\mu H^2 \end{bmatrix} \cdot \begin{pmatrix} 0 \\ 0 \\ 1 \end{pmatrix}.$$

$$= \begin{pmatrix} 0 \\ 0 \\ -\dfrac{1}{2}\mu H^2 \end{pmatrix} = \begin{pmatrix} 0 \\ 0 \\ -\dfrac{1}{2}BH \end{pmatrix}$$

(6.13).

The flux density between the lamination can be neglected. Equation 6.13 describes a force, which is perpendicular to the surface towards the lamination itself. The force density is about 320 N/m^2 for instance for a flux density in iron of 2 T and a relative permeability of 5000. This means a vibration amplitude of:

$$\Delta d = d \cdot \frac{F'}{E} = 0.5 \cdot 10^{-3}\,\mathrm{m} \cdot \frac{2 \cdot 320\,\dfrac{\mathrm{N}}{\mathrm{m}^2}}{2.1 \cdot 10^{11}\,\dfrac{\mathrm{N}}{\mathrm{m}^2}} = 1.5 \cdot 10^{-12}\,\mathrm{m} \cdot$$

The Eigen-frequency of this vibration can be calculated acc. equation 6.14:

$$f_1 = \sqrt{\frac{E}{\rho}} \cdot \frac{1}{d}.$$

(6.14)

This Eigen-frequency is for instance for a steel lamination of 0.5 mm at 1.0 MHz. Much lower frequencies as known from transformers are possible, if the lamination is not free and symmetrically supported over the complete surface. Often the vibrating surfaces are separated from the outer machine housing by cooling air. In these cases the cooling air and mechanical design elements will transmit the noise to the housing surface. Of course the housing will not vibrate in the same way as the inner source of noise. A transfer function needs to be determined. The determination of a proper transfer function is a serious challenge within noise calculation. The transmission through structure mechanical design elements can be determined by numerical field calculation, which would model the machine inclusive housing and auxiliaries like air intakes or fan caps in detail. Additionally the transmission through air according to equation 6.15 needs to be calculated:

$$\nabla^2 \vec{x} - \frac{1}{c^2} \cdot \frac{\partial^2 \vec{x}}{\partial t^2} = 0,$$

(6.15)

with space position $\vec{x}$, sonic speed c.

6.4 Influence of Pulse Frequency and Converter Operation on Noise Phenomena

Neither the calculation of the structure mechanic transmission nor the transmission through air will be done within this work.

6.4 Influence of Pulse Frequency and Converter Operation on Noise Phenomena

Electro magnetic forces have in a reasonable designed 2- or 4-pole electrical machine often less influence on the noise pressure level than aerodynamic modifications. The influence on the noise, which is caused for instance by different load conditions, is often not more than 3 dB. A modified fan may increase the noise directly by more than 6 dB, if the design change is not implemented carefully.

The significance of magnetic forces changes completely, if a motor is driven by a converter. Frequencies of several 100 Hz up to some kHz are caused directly by the terminal supply of the converter voltage in the stator flux density. Additionally the frequency of rotor or stator slot harmonics is no longer constant, but varies with the motor speed. Fig. 6.11 gives an example of the different time harmonics for the flux density in the air-gap of a 4-pole asynchronous machine, which is operated by a three-level converter. Even though the flux density and the resulting radial force amplitude, which is caused with the pulse frequency of the converter, are relatively small in their amplitude, the mode is equivalent to the mode of the fundamental. The resulting frequency would be $f_{force} = f_{pulse} \pm f_{fundamental}$. Low mode numbers with frequencies around 1 kHz are always a possible source for noise problems. The fundamental harmonic is 50Hz. The pulse frequency is the 19th harmonic or 950 Hz. The 4th Eigen-mode of the asynchronous machine has been found at 1500 Hz. It can be expected, that the stator vibration will not be excited by the converter pulse frequency, but some auxiliary equipment like terminal box may have to be looked closer at. A two-level converter with a higher pulse frequency is investigated in fig. 6.12. The pulse frequency is 1750 Hz. This frequency would again be far enough from the frequency of the 4th Eigen-mode, but smaller asynchronous machines with higher Eigen-frequencies have been excited by the interaction between the fundamental frequency and the pulse frequency of the converter.

Larger radial forces of about 1500 Hz can be found in fig. 6.13 due to the interaction of the fundamental with slot harmonics. Even though the force modes do not coincide with the mechanical mode at 1500 Hz, some mechanical asymmetries may allow the excitation of the 4th mechanical mode by the existing force modes. Due to the variable speed the modulation of the fundamental field with the rotor and stator slots should lead to an increased noise level of the 4th mode, if the forces act with the resonance frequency.

An overview of the magnetic forces in the air-gap for a two-level converter supply is given in fig. 6.14. Indeed the difference between the converter supplies is marginal in a direct comparison of the force densities. Converter imposed harmonics become only important, if the motor structure acts like an amplifier.

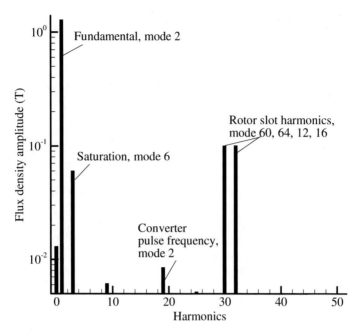

Fig. 6.11 Radial flux density amplitudes for frequency harmonics of the positive air-gap travelling waves, operation with a three-level converter

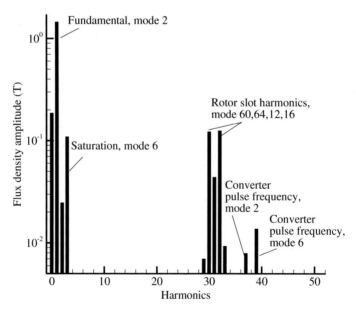

Fig. 6.12 Radial flux density amplitudes for frequency harmonics of the positive air-gap travelling waves, operation with a two-level converter

6.4 Influence of Pulse Frequency and Converter Operation on Noise Phenomena

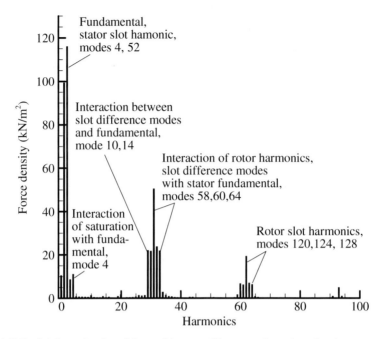

Fig. 6.13 Radial force density of the positive travelling wave for a three-level converter

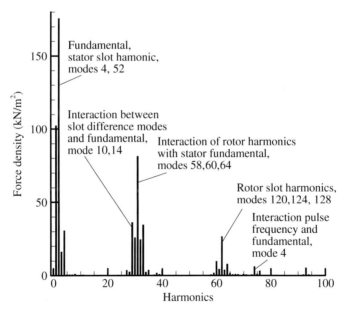

Fig. 6.14 Radial force density component of the positive travelling wave for a two-level converter on the stator surface

It is evident, that magnetic noise of converter driven machines can have many different roots. Main components within the air-gap force, which is acting on the stator, are summarized according table 6.3. Only those components are shown, which have been found in the investigated asynchronous machines to be dominant. Additional harmonics will occur for instance due to the non-sinusoidal distribution of the current around the machine circumferential. The more harmonics occur during sinusoidal operation, the more must be considered during converter operation as well.

Table 6.3 Space modes and frequencies, which have been dominant within the investigated asynchronous machines of m phases and p pole-pairs

Space mode	Frequency stator	Frequency rotor	Amplitude (%)	Reason
$2p$	$2\cdot f_{el}$	0	100	Fundamental field
$2p$	$f_{el}+f_{converter}$	$f_{pulse}-f_{el}$	2	Fundamental with converter harmonic
$3p+p$	$4\cdot f_{el}$	$2\cdot f_{el}$	14	Saturation with fundamental
$p+v$	$2\cdot f_{el}$	$2ma\cdot f_{el}$	14	$v=(2ma+1)\cdot p\,;\ a=0,\pm1,\pm2...$
v_1+v_2	$2\cdot f_{el}$	$2m(a_1+a_2)\cdot f_{el}$	<2	$v=(2ma+1)\cdot p\,;\ a=0,\pm1,\pm2...$
$v_1+Z_2-Z_1\pm p$	$(Z_2/p)\cdot f_{el}$	$(Z_1/p-2ma)\cdot f_{el}$	<1	Modulation of stator and rotor slot harmonics with fundamental and stator harmonic
Z_2-Z_1-p-p	$(Z_2/p-1)\cdot f_{el}$	$(Z_1/p-1)\cdot f_{el}$	15	Modulation of stator and rotor slot harmonics with fundamental and stator harmonic
Z_2-Z_1-p+p	$(Z_2/p)\cdot f_{el}$	$(Z_1/p)\cdot f_{el}$	27	Fundamental modulated with stator as well as rotor slot harmonics and with the fundamental
Z_2-Z_1+p+p	$(Z_2/p+1)\cdot f_{el}$	$(Z_1/p+1)\cdot f_{el}$	19	Fundamental modulated with stator as well as rotor slot harmonics and with the fundamental

Table 6.3 shows only few of the force harmonics. The focus on the chosen harmonics is of course the amplitude, the converter influence and a frequency range, which fits to the Eigen-frequencies of the first mechanical modes. Eccentricities can cause odd modes with amplitude of about 5 % of the fundamental harmonic force, but the frequencies of the forces are mainly two times the fundamental.

The amount of possible frequencies and modes according fig. 6.10 has to be taken into account, if an evaluation needs to be done, whether high noise levels must be expected. It is evident, that pulse frequencies, which will excite mechanical Eigen-modes by the fundamental space harmonic, must be avoided. The influence of the stator and rotor slots is less clear. The different space harmonics of the force density have been multiplied according to the influence of the mode r with

6.4 Influence of Pulse Frequency and Converter Operation on Noise Phenomena

$1/(r^2-1)^2$ and are shown in fig. 6.15. The considered 4-pole motor has 48 slots in the stator and 62 slots in the rotor. The motor will be operated in a speed range between 1500 rpm- 3000 rpm. The fundamental space harmonics and the harmonics due to the eccentricity will not cause dominant noise levels, because the frequency range of the forces is below 200 Hz. The interaction of the 5th stator harmonic in the flux density with the slot harmonics or the slot harmonics itself will at certain speeds excite the stator core to vibrations. Forces with 1500 Hz will act either on the stator surface or on the rotor surface.

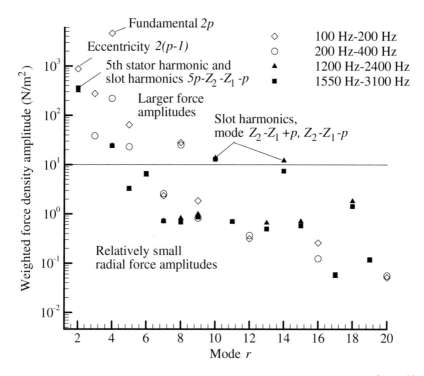

Fig. 6.15 Force density amplitudes of stator and rotor forces weighted with $1/(r^2-1)^2$

All frequencies in fig. 6.15 have been determined based on a converter supply with sinusoidal voltage characteristic. This can be achieved by a multi-level converter or a filter between the converter and the motor. More harmonic content can be expected in the voltage supply from other converter designs. Fig. 6.16 shows the Fourier-analysis of a three-level voltage source converter at 105 Hz fundamental frequency. The implemented control strategy gets its pulse muster from predefined tables. It is often the target to reduce converter losses by low pulse frequencies. Higher frequencies would cause more switching losses and would require decreasing the converter output. Two different pre-calculated pulse patterns have been chosen in fig. 6.16 a) and 6.16 b) in order to get an imagination on the influence between the chosen pulse frequency and the motor noise.

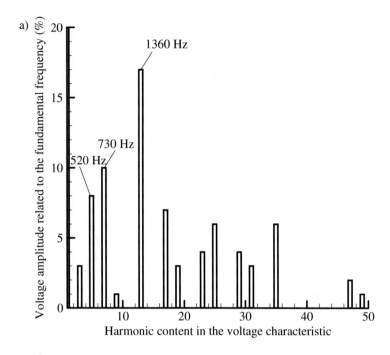

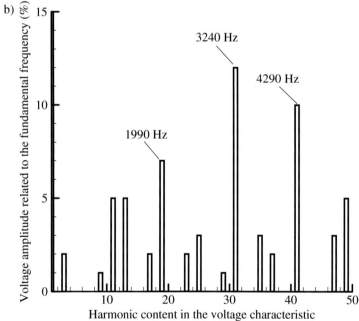

Fig. 6.16 Harmonic content of the motor terminal voltage supplied from a three-level frequency converter; a) low pulse frequency, b) higher pulse frequency

6.4 Influence of Pulse Frequency and Converter Operation on Noise Phenomena

The voltage characteristic in fig. 6.16 a) has main frequency components with 520 Hz, 730 Hz and 1360 Hz. Especially the 1360 Hz component together with the 105 Hz fundamental will cause force densities with 1465 Hz, which is very close to the first mechanical Eigen-mode of the measured machine. These converter caused noises are given for the fourth mode in fig. 6.17. A noise pressure level of 86 dB(A) has been measured in this operation mode. The voltage characteristic in fig. 6.16 b) shows a dominant frequency component of 3240 Hz. This is far beyond the Eigen-frequencies of the low order stator Eigen-modes. The measured noise has been reduced by 9 dB(A) down to 77 dB(A).

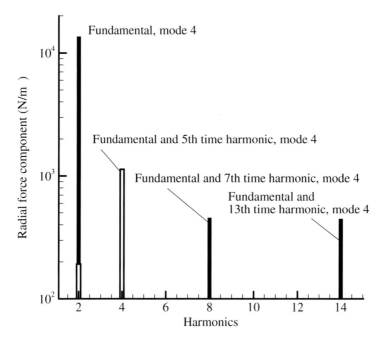

Fig. 6.17 Fourth mode of the radial force density of the positive travelling wave for a low saturated 4-pole 1400 kW motor which is supplied by a three-level converter

The following rules may be applied for low noise converter driven machines:

- Coincidences between the frequencies of the harmonic content in the stator voltage characteristic plus the fundamental frequency and any Eigen-frequency of the mechanical core vibration mode should be avoided. Especially the mechanical frequency from the Eigen-mode of $2p$ must not be equivalent to the pulse frequency plus the frequency of the fundamental,
- The number of rotor slots multiplied with the rotation speed frequency should be below or above the first four stator Eigen-modes,
- Application of water cooling to reduce ventilation caused noise,
- The number of stator and rotor slots has to be chosen with focus on minimum stator higher harmonic currents,

- The lamination package must be packed dense to reduce lamination vibrations,
- Low air-gap flux density reduces acting forces,
- Optimization between core modes, Eigen-frequencies and pole pairs vs. harmonic voltage content.

Additionally investigations have been initiated to operate with randomly modulated carriers, see [6.12]. This strategy may be successful, if an adjustment of the converter to the individual motor Eigen-modes could be standardized.

Chapter 7
Converter Caused Shaft Voltages

7.1 Overview of Different Shaft Voltage Types

Parasitic effects caused by static frequency converters, like electromagnetic inter-ference (EMI), may endanger the reliable operation of electrical machines up to 500 MVA. They can cause shaft voltages up to 300 Vpp, if a non optimized converter design is chosen. Shaft voltages occur in converter fed drives due to different effects [7.1-7.9]. With the application of power electronic devices, new types of shaft voltage sources occur [7.1, 7.2, 7.8, 7.9]. In general shaft voltages due to power electronics are either capacitive or inductive. An overview of the converter imposed types of shaft voltages for motor sizes over 200 kW is given in table 7.1.

Table 7.1 Types of converter caused shaft voltages

Type of shaft voltage	Main reason	Voltage peaks	Current peaks	Frequency
Circulating flux in-duced voltages [7.10]	Capacitive currents within the stator winding cause a circulating flux	up to 300 Vpp	>70 A	100 kHz to 1000 kHz
Capacitive imposed shaft voltages [7.3]	Frequency converter feeds rotor wind-ings in static excitations or double fed induction machines	up to 80 Vpp		ca. 200 kHz
Capacitive imposed shaft voltages [7.1]	Frequency converter feeds stator winding, which is capacitive coupled to the rotor	up to 50 Vpp	>20 A	100 kHz to 1000 kHz
Capacitive imposed shaft voltages [7.6]	Ineffective stator grounding may lead to a floating bearing potential	less than the voltage potential be-tween motor and inverter frame <100 V		

Three types of shaft voltages dominate in industry. The dominance depends upon the machine size and design. If the rotor winding in static excited synchron-ous machines or double fed induction machines is fed by a converter, the capaci-tive coupling between winding and shaft can be seen. Capacitive coupled shaft voltages occur in converter fed stator windings in case of machines with small air-gaps. The third type is mainly dominant in larger machine sizes of stator fed windings. A part of the high frequency currents is decoupled from the stator-winding flowing to ground. The unidirectional current flow causes a circulating flux tangentially in the stator joke. This flux induces a voltage.

7.2 Circumferential Flux and Capacitive Imposed Shaft Voltages

In case of a converter fed stator the capacitive coupling to the rotor is relatively small for machine sizes over 500 kW due to relative large air-gaps. The dominant shaft voltage amplitudes are induced by circulating fluxes. The circulating flux effect is shown in the following in case of a large start up converter for a turbo-generator.

Recent developments in power generation equipment show an increasing growth in the installed unit power of gas turbines up to 340 MW. Whereas steam turbines are capable to run up independently, gas turbines have to be started and driven until they turn fast enough to allow ignition and self-sustaining operation. Ignition may not be possible up to 80 % of nominal speed. The turbine is often started by the application of the turbo-generator as motor. Electrical power is supplied to the generator-motor by a variable-frequency starting device like a static frequency converter (SFC). A typical start arrangement is given in fig. 7.1.

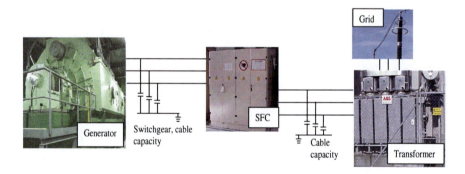

Fig. 7.1 Electrical power plant equipment for starting the gas turbine

The SFC has to deliver considerable amounts of power, e.g. 5-15 MW for a typical large gas turbine. The SFC is disconnected from the generator after running up the gas-turbine to ignition and self-sustained operation. The generator is consequently used as generator then. Usually this start-up operation with active SFC takes only a couple of minutes, e.g. seven minutes, which is quite a short period of time. Because of this short period some effects and phenomena during start-up were not discovered for a long time [7.10].

7.2 Circumferential Flux and Capacitive Imposed Shaft Voltages

Due to modern use of gas-turbine plants to generate peak-load, turbines must be switched on and off several times per day. In-between times of full operation periods of active SFC-operation for maintenance purposes exist. Therefore shaft voltages during starts with static frequency conversion have to be cared for more and more.

If this kind of circulating flux induced shaft voltage reaches certain values, different effects can be observed at a typical power plant. Strong electromagnetic interferences occur at sensors, video screens and distributed control equipment (DCS) at a typical 300 MVA gas-turbine plant during SFC-operation. The powerful SFC and the close proximity of control equipment to the SFC has been the main reason for these disturbances. The grounding and shielding of systems are improved to reduce the interference to acceptable levels, which appeared to be quite costly. Furthermore currents can reach amplitudes, which are sufficient to blow the fuse of the non-driven end (NDE) shaft-grounding module, "RC-module", in fig. 7.2. A typical grounding arrangement is shown in fig. 7.2. The shaft has its direct grounding at the driven-end (DE) of the generator. At NDE the shaft is additionally grounded by a mainly capacitive element. No low-frequency current can flow, but high-frequency voltage peaks, which are typically caused by the thyristors of the excitation system, are strongly damped [7.11].

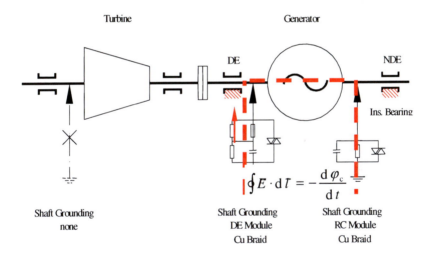

Fig. 7.2 Grounding scheme of shaft train with RC-module

Measurements have revealed, that voltage peaks with abnormally high amplitudes, fig. 7.3, during active SFC-operation are the cause of currents of more than 70 A, which have even blown the fuse. The voltage pattern is in its timely distribution of peaks independent of the turning speed of the turbine. Every 6.7 ms a strong peak appears, followed by a much lower peak after 3.3 ms. It is evident that the rectifier at the input-side of the SFC has to be considered as the primary source. A source for shaft voltages in form of the converter has been found on asynchronous machines also. Some publications are related to experience with converter fed asynchronous machines [7.7-7.9], which have an inductively coupled shaft voltage. The currents at DE and NDE must be the same in amplitude, but with different direction. Synchronized measurements on the gas turbine set at the DE-side and the NDE-side showed, that a positive pulse at the NDE-side coincides with a negative pulse at the DE-side and vice versa. The picture of a closed current loop arises with the current being induced by changing magnetic fields in the stator core. This hypothesis is supported by the observation that the pulses at NDE have low-impedance sources. Current strengths of up to 70 A were measured, which could not be explained by capacitive coupling. These kinds of shaft voltages are clearly inductively caused with the SFC as main root cause.

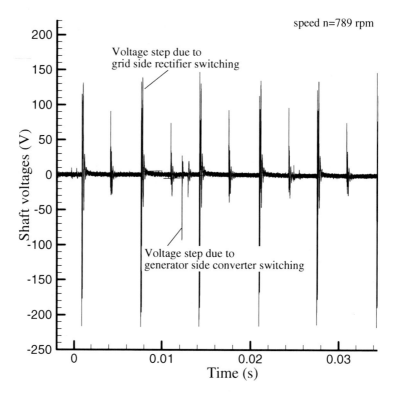

Fig. 7.3 Shaft voltage peaks at NDE during turbine run up

7.2 Circumferential Flux and Capacitive Imposed Shaft Voltages

A typical electrical design of a converter fed plant drive is given in fig.7.4.

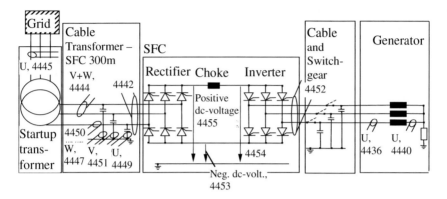

Fig. 7.4 Simplified schematic of main electrical plant components during start-up

Fig. 7.4 reveals a strong asymmetry concerning the arrangement of the main choke between the rectifier and converter. The static frequency converter consists of two thyristor bridges. It is a current-type converter. Between both bridges a d.c. circuit exists with one choke in the positive branch. The negative branch connects both bridges directly. The bridge at the grid side is used to adjust the voltage level

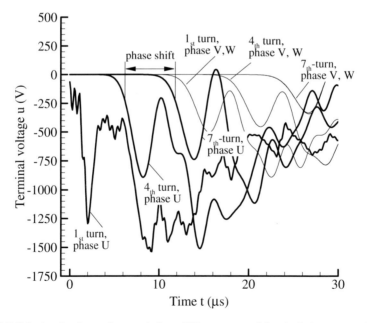

Fig. 7.5 Calculated voltage characteristic at different turns of the single bar stator winding

of the d.c. circuit and hence the d.c. current. It works as a rectifier. During one period of 20 ms each of the 6 thyristors is switched on and off. Three of them are connected to the positive branch with the choke, the other three are directly connected to the output rectifier bridge. The bridge at the machine side works as a converter. At lower speeds it operates in block mode, afterwards it works as LCI inverter. If a thyristor is switched on, a voltage surge is caused at the generator terminals. The surge acts on the stator winding with its capacitive and inductive elements like a traveling wave. The phenomenon has been calculated and is shown in fig. 7.5. Indeed fig. 7.5 reveals a phase difference in the voltage towards ground of about 5-10 µs between three turns of the same winding. A similar time shift can be seen between the stator terminal and stator neutral currents. The time constants of the shaft voltage pulses and the difference current between the stator terminal-current and neutral-current are fairly similar. It will cause a circular magnetic flux in the stator core and hence induce the shaft-voltage pulses. In order to calculate the amplitude of the induced shaft voltage this traveling wave phenomena has to be considered. An overview of the appropriate calculation model is given in fig. 7.6.

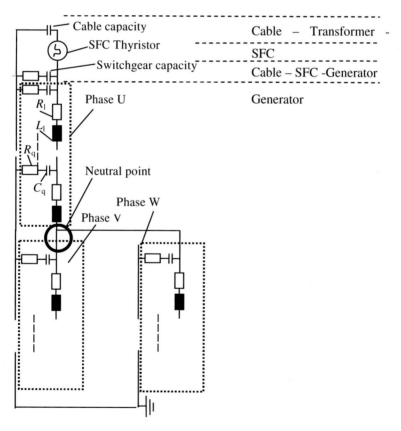

Fig. 7.6 Circuit diagram of the turbo-generator connected to a static frequency converter

7.2 Circumferential Flux and Capacitive Imposed Shaft Voltages

Several approaches are available to solve this kind of circuit model [7.12-7.14]. The individual circuit parameter can are derived from machine and plant quantities. Especially the individual influence of the short time constants of the transients have to be taken into account. Based on table 7.2 and 7.3 a resistance increase within a multi-strand stator winding can be derived for an exponential current increase. The estimation is based on the application of the skin time [7.15]. The skin time is defined according to equation 7.1.

$$R_{ac} = \frac{\delta_{T,bar}}{T} \cdot \frac{d}{\delta_s} \cdot R_{dc} \, . \tag{7.1}$$

Whereas table 7.2 shows the loss increase within the strand, which is current leading, table 7.3 gives an overview about the losses within the adjacent strands.

Table 7.2 Equivalent resistance and skin time for different geometries and inherent exponential $(1 - \exp(-t/T))$ current flow

Strand depth d (mm)	1.5		
Time constant (μs)	10	100	1000
δ_s (mm)	1.5	1.5	1.5
δ_T (μs)	33	99	980
b/l·R_{acsT} (mΩ)	0.040	0.012	0.012
Strand depth d (mm)	2		
Time constant (μs)	10	100	1000
δ_s (mm)	1.55	1.82	2.0
δ_T (μs)	49	99	990
b/l·R_{acsT} (mΩ)	0.057	0.010	0.009
Strand depth d (mm)	3		
Time constant (μs)	10	100	1000
δ_s (mm)	1.7	3	3
δ_T (μs)	90	150	1000
b/l·R_{acsT} (mΩ)	0.095	0.009	0.006

Table 7.3 Equivalent resistance and skin time for different geometries and current flow in adjacent strands

Strand depth d (mm)	2		
Time constant (μs)	10	100	1000
δ_s (mm)	1.28	1.36	1.33
δ_T (μs)	111	32	4
b/l·R_{acsT} (mΩ)	0.155	0.004	0.00006

116   7 Converter Caused Shaft Voltages

The skin time for a bar with i strands can be derived by the following equation 7.2, if eddy currents are dominant:

$$\delta_{T,bar} = \delta_{T,1} + \frac{\delta_{T,2}}{i} \cdot \sum_{k=1}^{i} (k-1)^2 \qquad (7.2)$$

The skin time $\delta_{T,1}$ is given in table 7.2, whereas the skin time $\delta_{T,2}$ is given in table 7.3. The additional machine parameters are calculated as follows:

$$L_1 = \frac{1}{n} \cdot \left(k_i \cdot L_{\sigma,bar} + L_{\sigma,s} \right), \qquad (7.3)$$

$$C_q = \frac{1}{a \cdot n} \cdot C_a \qquad (7.4)$$

$$R_q = \frac{1}{a \cdot n \cdot \gamma_{Fe}} \cdot \left(l_{Fe,core} + 0.5 \cdot l_{Fe,tooth} \right) \cdot 2\pi \cdot r_m \cdot l_{Fe}, \qquad (7.5)$$

with number of chain conductor elements per phase and per parallel circuit n, major time constant T, in the given example 10μs, depth of one strand of the bar d, resistance of one stator phase R_{dc}, skin depth δ, skin time $\delta_{T,bar}$, leakage inductance $L_{\sigma,bar}$, component along the bar cross section, leakage inductance $L_{\sigma,s}$, component of the slot-opening and air-gap, capacitance of one stator phase C_a, number of parallel circuits a, mean radius of stator core r_m, effective axial iron length l_{Fe}, iron length between stator outer diameter and slot-ground $l_{Fe,core}$, length of a stator tooth $l_{Fe,tooth}$, conductivity of iron γ_{Fe}.

The right choice of the resistance along the bars, the inductance along the bars and the capacities between the bars to the stator core is essential for the right prediction of the induced shaft voltage. The consideration of stator lamination is done only on a very basic level. Indeed it is a complex three-dimensional field problem, which would have to consider the capacitive coupling between each lamination and the eddy current phenomena within the laminations itself. After the laminations the machine design foresees a short circuited cage elastically coupled over massive iron rings to the housing. An improved approach for the laminations neglecting the capacitive effect is given in [7.1]. The resistance of the core is an area for further investigations [7.16].

The high frequency model, which has been described, is applied to the running up of a 300 MVA air-cooled generator. The generator is coupled to a gas turbine and is used as a motor to start the shaft train. A plant has been chosen, where the very particular arrangement supports the generation of shaft voltages. Any calculation model can only be regarded as valid, if it has been verified by measurements. The measured thyristor voltage has been impressed in the model and the calculated stator terminal current is compared with measurements in fig. 7.7. The measured and calculated currents fit well for the first 40 μs after the thyristor switching. The measured current is 131 A, whereas the calculated amplitude is about 20% smaller. This deviation is within the amount of the measured shaft

7.2 Circumferential Flux and Capacitive Imposed Shaft Voltages

current and voltage variation. Indeed the waveform was found to be similar independent upon its magnitude and the site, where the measurements were taken. After a first period the dominant frequency decreases and the applied calculation model needs to be adjusted. This has not been done due to the primary interest in the maximum or minimum peaks.

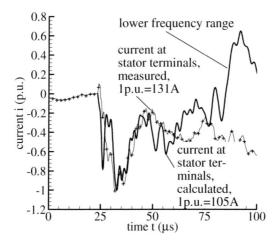

Fig. 7.7 Comparison between calculated and measured stator terminal current at 798 rpm, different rated currents have been chosen in order to have a better comparison of the waveform

Indeed an exact determination of the right induced bearing voltage and the difference axial current would request an appropriate knowledge about the mutual inductance between the shaft and the capacitive decoupled current path. An approach for the total stator ground current is given in [7.1]. Again the capacitive current path is an assumption of radial current flow within the lamination, equation 7.6 and 7.7:

$$L_g = \frac{\mu \cdot N_{Fe}}{2 \cdot \pi} \cdot \ln\left(\frac{d_{se}}{d_{si} + 2 \cdot h_s}\right) \cdot \frac{\delta_s}{2} \quad (7.6)$$

$$\delta_s = \sqrt{\frac{2}{2\pi f \gamma \mu}}, \quad (7.7)$$

with mutual inductance between stator ground current and shaft L_g, Number of stator laminations N_{Fe}, stator outer diameter d_{se}, bore diameter d_{si}, slot-height h_s.

Considering a shaft grounding acc. fig. 7.2 a much simpler approach is possible. In real applications the current flow in the shaft will reach similar amplitudes as the difference current. The capacitor is nearly short circuited due to the high frequency.

Machines with a relatively large capacitive coupling between stator winding and rotor have often a remarkable capacitive coupled shaft voltage beside the circulating flux phenomenon. This is more typical for asynchronous machines than for synchronous machines. The shaft voltage level depends mainly upon the individual distribution of capacities. An overview of the dominating capacities is given in fig. 7.8. The stator winding is coupled via the air-gap to the rotor. Of course this capacity is smaller than the coupling to the stator frame. Capacities occur between the rotor and the bearings as well as between the rotor via the air-gap to the stator core in series to the stator winding rotor coupling.

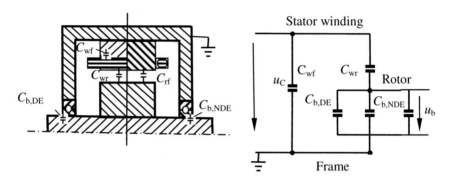

Fig. 7.8 Capacitive machine model, drawings acc. [7.1]

A sinusoidal three-phase voltage system is coupled to the rotor as well, but the rotor would be similar to a neutral point, which is created by capacities. In case of a common mode voltage the situation is different. The shaft voltage over the bearing u_b is related to the common mode voltage u_{CM} by the following relationship:

$$\frac{u_b}{u_{CM}} = \frac{C_{wr}}{C_{wr} + C_{rf} + 2 \cdot C_b} \cdot \qquad (7.7)$$

Whereas the voltage is properly defined by equation 7.7, the current is not. Indeed the current is a result of different phenomena [7.1].

In case of a slip ring supply for the rotor winding, a capacitive shaft voltage dominates any other kind of shaft voltage in general, if the shaft is not properly grounded. This case is evident for slip ring turbo-generators as well as for slip ring asynchronous machines. Main capacities are given in fig. 7.9.

The strong capacitive coupling is on such a level, that any machine with direct converter supply to the rotor has to have proper rotor grounding. The voltage can be calculated according equation 7.8:

$$\frac{u_b}{u_{CM2}} = \frac{C_{rw2}}{C_{rw2} + C_{rf} + 2 \cdot C_b} \cdot \qquad (7.8)$$

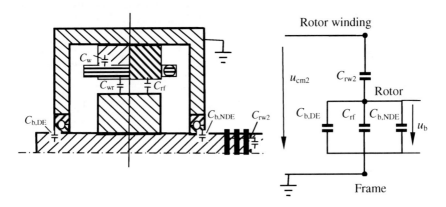

Fig. 7.9 Capacities within a slip ring asynchronous machine

7.3 Measurements of Current Path and Voltage Transients

Shaft voltages of several hundred volts occur during the operation of turbo-generators or other large electrical machines with static frequency converters, if certain conditions are fulfilled. These conditions can be elaborated by measurements. The complete system including the transformer, cables, converter and the electrical machine has to be regarded.

Currents of high frequencies up to several 100 kHz can not be measured by normal current transformers. Other measurement techniques are necessary.

High voltage and frequency applications require a special handling. Although the application of measurement shunts is a reliable means for 50 Hz or 100 kHz for low voltage applications, it gets much more complicated for high voltage, high frequency applications. Current transformers with iron cores can not be used due to inherent eddy currents. Therefore Rogowski coils without iron are used, fig. 7.10. These coils consist of a thin copper wire, which is wound around non-magnetic, non-conductive material.

A systematic measurement approach is shown on the start up system of a gas turbine turbo-generator. All involved electromagnetic elements during the start up of the shaft train are shown in fig. 7.11. The measurement arrangement is equipped with ten Rogowski coils and three voltage probes in the SFC. They are numbered in fig. 7.11 and the signals will be referred to by these numbers.

In order to understand the shaft voltage phenomenon properly it is essential to know the right current path and the source of the high frequency voltage drop. In principle two elements within the arrangement of fig. 7.11 could cause the voltage drop. One source could be the inductances within the transformer. During the switching of the rectifier thyristors one line of the transformer is switched off and the transformer inductances may induce a high voltage peak. The other possible

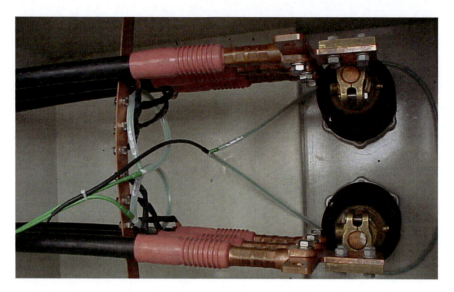

Fig. 7.10 Current measurement with Rogowski coils at the start transformer

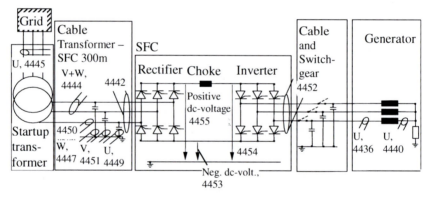

Fig. 7.11 Plant arrangement and measurement points MP 4436-4455 for the start up of the power train

element could be the thyristor bridge itself. During the commutation phase the thyristors cause the voltage peaks within the d.c. link. The thyristor commutates in two steps. In a first step the thyristor is connected to the d.c. link in parallel to the already connected thyristor causing a voltage step. Then the current passes over to the newly connected one until the current in the other thyristor becomes negative and this thyristor is disconnected. The disconnection of the thyristor causes a second voltage step. Additionally it might be possible that some outside component, which is fed in by the start up transformer, causes the peaks.

The last assumption is not the case. This is shown in fig. 7.12. On the high voltage side of the start up transformer the peaks are smaller than on the low voltage side. In the following, the high frequency current path is investigated.

7.3 Measurements of Current Path and Voltage Transients

Depending upon the voltage or current source the current path is different. The complete path could be described by the following six measurement points:

- Current through the transformer (MP 4445, MP 4450)
- Current into the SFC (MP 4442)
- Current out of the SFC (MP 4452)
- Current in the generator (MP 4436)
- Current at generator neutral (MP 4440)
- Current in the transformer – SFC cable grounding (MP 4447, MP 4449, MP 4451)

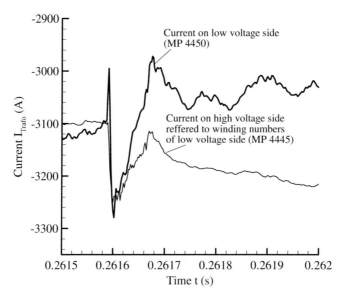

Fig. 7.12 Parasitic currents within the start up transformer during thyristor switching on low and high voltage side

If the transformer inductances are the source, the path should be as follows:

The current should go through the transformer inductance of the switched off line, then through another line to the SFC, then out of the SFC in the generator. The current, which is going in the generator, should not come out directly at the generator neutral, but over the cable grounding to the transformer again.

If the thyristors are the source, the path should be different:

The current should go through the commutated thyristor, then out of the SFC to the generator. The current, which is going in the generator, should not come out directly at the generator neutral, but over the cable grounding to the thyristor again.

Fig. 7.13 shows the transformer current and the shaft voltage during the commutation period of the thyristor bridge. The current within the start up transformer reduces from one level to another during the thyristor switching. A direct relation to the shaft voltage can not be derived. A clear relationship between the shaft voltages and the transformer currents should exist in case of the transformer as voltage source. Fig. 7.14 shows the summation of all three transformer phases.

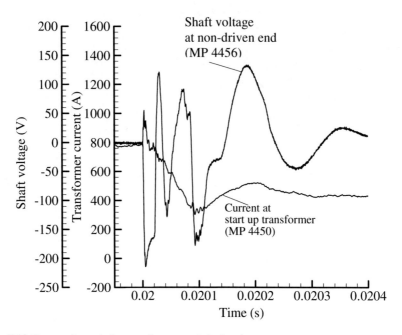

Fig. 7.13 Current through the transformer and shaft voltage

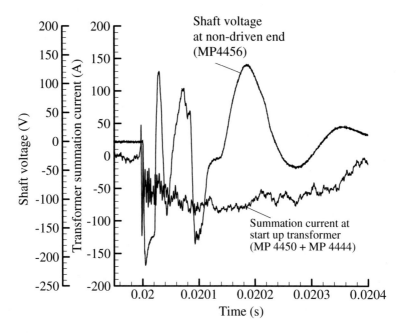

Fig. 7.14 Relationship between the summation current of all transformer phases and shaft voltage

7.3 Measurements of Current Path and Voltage Transients

Also no relationship between the shaft voltage and the summation current can be found in fig. 7.14. Therefore the transformer can be excluded from the potential main cause of the voltages. Fig. 7.15 shows the summation of all currents into the SFC and out of it. Indeed the circuit of this summation current is closed by the grounding system. Two main results can be drawn from fig. 7.15. On the one hand the current flow into the SFC is equivalent to the current flow out of the SFC. This means that there are no connections between the SFC and the grounding system, which would allow for a direct or capacitive current flow to the grounding system. On the other hand the second result, which can be seen by a comparison with the shaft voltage, is even more interesting. The shape of the shaft voltage is very similar to the shape of the summation current. Whereas the transformer as source of the shaft voltage phenomenon could be excluded, the SFC seems to be the real source.

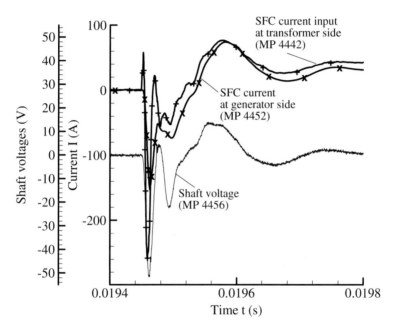

Fig. 7.15 Current into the SFC and out of the SFC

It has been shown how the shaft voltage is induced by a circulating flux flow through the generator core, [7.10]. The circulating flux flow can only occur in case of a difference between the current in the positive and negative axial direction. Therefore the current of the stator terminals and the neutrals is measured, fig. 7.16.

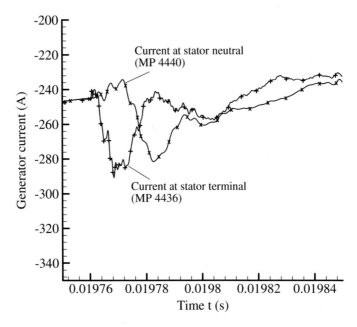

Fig. 7.16 Current into the generator terminals and at the generator neutrals

Indeed fig. 7.16 reveals a phase difference of about 10-15 µs between the current at the terminals and at the neutral. The time constants of the shaft voltages and the difference current between the stator terminals and neutrals are fairly similar. The theoretical explanation of this kind of shaft voltages is verified by these measurements.

Further understanding of the shaft voltage phenomenon is acquired by the currents through the grounding system, fig.7.17. During the commutation phase a high frequency resonant current occurs between both involved phases on the rectifier bridge of the SFC. The currents are strongly damped after the commutation phase comes to its end and one thyristor starts to interrupt the short circuit of both phases.

Whereas the high frequency Eigen-mode of the circuit cable – thyristor – cable – grounding and back to the cable is dominant in each of both concerned phases, the summation of the phases reveals the relationship with the shaft voltage, fig. 7.18.

The overall result of the investigation of the current path is given in fig. 7.19. Only the main involved components are drawn.

Three main current circuits are dominant at different localizations and different times. Locally limited high frequent oscillations with about 100 kHz are generated in the elements before the thyristor bridge as described above. At the same time the first current peak reaches the generator and causes a travelling wave in the generator. After the commutation phase of the thyristors the third current circuit becomes active. The cable capacitance, the switchgear capacitance, the generator capacitance and inductance built a resonant circuit, which creates a current oscillation with a frequency of about 10 kHz.

7.3 Measurements of Current Path and Voltage Transients 125

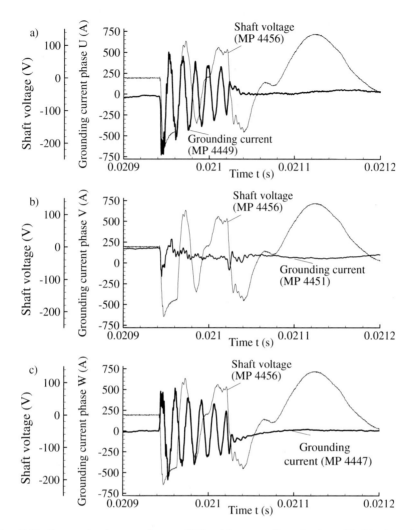

Fig. 7.17 Current in the transformer–SFC cable grounding a) phase U, b) phase V, c) phase W

In order to understand the influence of the converter better the voltage distribution in the converter is determined. The static frequency converter consists of three main parts. It has a rectifier bridge at the grid side and a converter bridge at the generator side. Both thyristor bridges are connected via a d.c. link. The d.c. link has on the positive side a choke, which reduces the current oscillations. The other link connects the rectifier bridge directly with the converter bridge. The voltage between this link and ground is given in fig. 7.20. A strong voltage drop of about 4kV is caused, if the link is switched from one thyristor to the other.

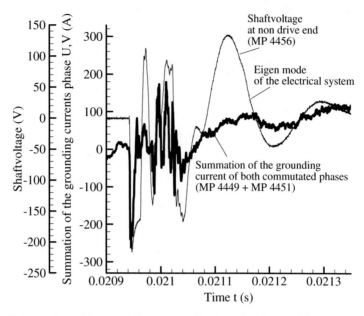

Fig. 7.18 Summation of the grounding currents in phase U and phase V

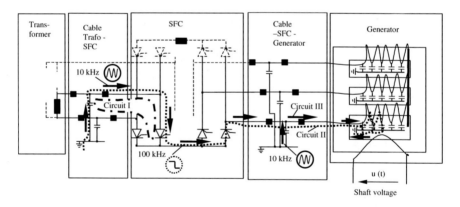

Fig. 7.19 Description of all three current paths of parasitic effects during thyristor switching

In-between the thyristors of the positive side are switched, but this causes only a small voltage drop. Fig. 7.21 shows a zoom of one of the main voltage drops. As soon as the second thyristor is fired, the d.c. link voltage reduces with a gradient of 160 V/µs. During the commutation phase the voltage level stays on an intermediate level between both transformer phase voltages and in the end one thyristor interrupts the current circuit.

7.3 Measurements of Current Path and Voltage Transients

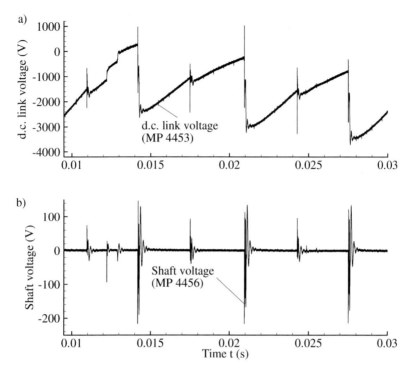

Fig. 7.20 a) voltage in the negative d.c. link and b) shaft voltage during LCI operation

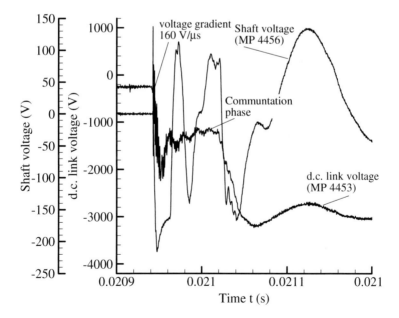

Fig. 7.21 Shaft voltage and voltage in the negative d.c. link during LCI operation

Whereas the shaft voltage is determined in the beginning by the voltage gradient, the system is in self-resonance after the commutation.

The situation is different in the positive link, fig. 7.22. Here a choke is included in the circuit. The voltage drop at the grid side is even higher in amplitude, up to 7 kV, than the one in the negative link and the frequency is twice as much. The voltage drop behind the choke at the generator side is much smaller due to the strong coupling to the generator. The shaft voltage during a commutation on the positive side of the bridge is nearly a factor of three smaller than on the negative side. The main reason for this is the strong damping by the choke, see fig. 7.23.

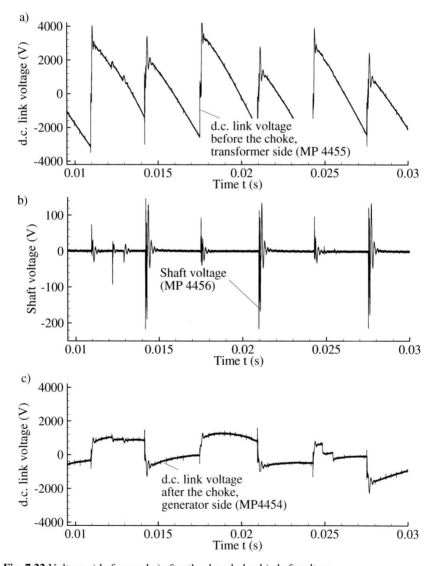

Fig. 7.22 Voltage a) before and c) after the d.c. choke, b) shaft voltage

7.4 Shaft Grounding and Converter Concepts

The voltage characteristic behind the choke increases about three times slower than the one in the circuit without a choke. This result is an evidence for the correlation between the shaft voltage and the gradient of the thyristor bridge voltage.

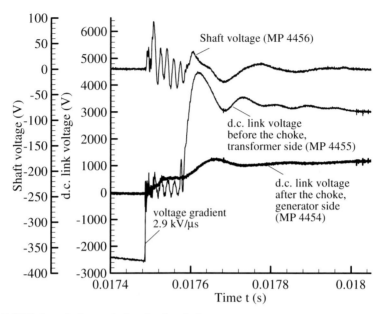

Fig. 7.23 Voltage before and after the d.c. choke

7.4 Shaft Grounding and Converter Concepts

Maximum allowable values for shaft voltages are determined by several different factors. It should be avoided that touching the shaft could harm anybody. Otherwise it would require a huge additional effort to secure the shaft in a way that no non-authorized personnel could touch the shaft. Maximum touch voltages are determined within different standards [7.17, 7.18]. Here it is necessary to distinguish between the voltage level of the shaft and the voltage the human being sees. These voltages vary within the range of application. No standardized limits exist so far for shaft voltages. Therefore values for cases, which could be transmitted to shaft voltages, are listed in table 7.4. Indeed shaft voltages differ that much that it would not be sufficient to determine just one limit for all kinds of voltages. Table 7.4 especially refers to 50 Hz or 60 Hz applications. Additionally the touch voltage will be strongly different between only capacitive coupled voltages with low danger for human beings and those, which are also inductively caused. These inductive sources are capable of causing current amplitudes, which may result in severe harm to human beings. In line with the described standards it can be considered that there exists no direct major danger for healthy personnel from the electrical currents as long as the shaft voltage is lower than 125 Vrms or 177 Vpeak. An additional danger is of course that the personnel will not expect voltages on the shaft. Therefore it is necessary to inform in respect to this point.

Table 7.4 Maximum shaft voltage levels according to standards

Normal application	Allowable shaft voltage	Allowable touch voltage	Standard
Grounding in a.c. application	<125 Vrms	--	[7.17]
	>125 Vrms	<65 Vrms	[7.18]

Beside the limits, which are defined by the protection of human beings, an additional danger is seen for the bearings. Shaft voltages can cause electro-erosion of the bearing shells and shaft seals [7.11]. Whereas the generator bearings are insulated, the bearings of the turbine are usually not insulated. The limit due to non-insulated bearings can be derived from the work, which has been performed in [7.3, 7.11].

The authors mention a fairly conservative limit for non-insulated large sleeve bearings of 20 V. At the same time they measured that up to 40 V no breakdown occurs. This safety margin is quite high. During the long experience of shaft line commissioning a reliable value has been found to be 35 V. This value still leaves a margin of 5 V to the first measured breakdown. Of course even if a breakdown occurs from time to time, it is necessary to differ between a steady state operation and a transient situation for instance during start up.

Additionally no distortion to the measurement and protection equipment has been seen provided the prescribed limits are kept.

Different measures can be applied in order to reduce shaft voltages of converter drives. Modifications to the converter could be considered as well as appropriate shaft grounding concepts. Concepts for industrial drives and turbine generator sets are shown in fig. 7.24. The kind of shaft grounding depends strongly upon the machine size as well as on its application.

Low voltage drives of up to 1000 kW with random winding arrangements are supplied in general with IGBT converters. Even though their voltage surge is relatively steep, it is generally sufficient to insulate the NDE bearing.

In case of medium voltage converters the voltage surge at the terminals is about six to fifteen times larger. These large voltage surges dominantly influence the shaft voltage, which has its origins in a circulating flux around the shaft. A circulating current loop including the shaft would be the result and has to be interrupted. An example is given in fig. 7.24 b). The drive side of the motor is equipped with an insulated bearing. Additionally the coupling towards the load is insulated. In order to reduce capacitive caused shaft voltage phenomena, the non drive end is properly grounded.

Grounding concepts within large power plants are explained in [7.2, 7.3, 7.10]. These concepts are applied mainly on shaft arrangements for turbo-generators with slip ring excitation. Different concepts have to be applied in case of turbo-generators with single or double shaft end arrangements, see fig. 7.25. The concepts have to consider several kinds of shaft voltage phenomena, which could damage the bearings of the generator as well as of the turbine. Beside shaft voltages due to converter fed slip rings or stator windings, shaft voltages do occur due to magnetic asymmetries, electrostatic charge or axial flux.

7.4 Shaft Grounding and Converter Concepts

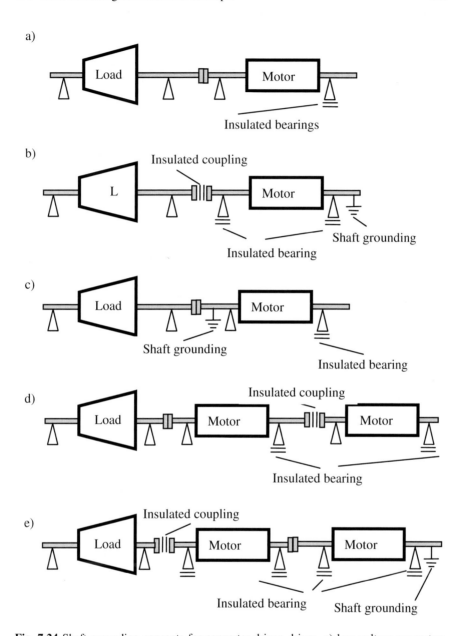

Fig. 7.24 Shaft grounding concepts for converter driven drives, a) low voltage converters with anti friction bearings, b) medium voltage converters or low voltage with sleeve bearings, c) medium voltage converters for small drive end shaft diameters, d) low voltage tandem drives, e) medium voltage tandem drives

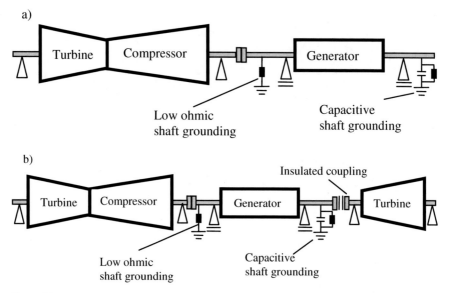

Fig. 7.25 Grounding concepts for turbo-generator sets with converter supplied excitation winding or stator winding during start up

Whereas the first two cases are proven shaft voltage sources, the axial flux causes only a d.c. shaft-voltage in case of flux and shaft grounding, which is equivalent to the arrangements of an uni-polar motor [7.19]. An uni-polar shaft-voltage is induced in the case, where the shaft carries some magnetic flux at the grounded positions.

Beside concrete measures to the machine arrangement, the converter design influences the shaft voltage amplitude as well. Within the chapter about converter type topologies a comparison between the terminal voltages of multi-level- and two-level converters has been given. The multi-level low voltage design together with a relatively high pulse frequency causes low levels of voltage peaks. The extremely drive friendly converter is dominant within the oil and gas industry with a special tradition for high availability. Basic rules can be established, which should be applied to keep shaft voltages below a value which in general does not cause damages to bearings. Large insulated sleeve bearings should not see more than 70 V in order to avoid malfunctions in instrumentation, non insulated sleeve bearings should see shaft voltages below 30-35 V in order to avoid breakdowns over the oil film. Rules in order to keep these limits are often kept within the fields of motor design, converter design and plant design. The rules are the result of several years of damage experience and research. Shaft voltage caused damages often do only occur, if several of the rules are broken at the same time. Additionally rules depend strongly on the kind of the dominant voltage type and therefore on the size of converter driven machine. Nevertheless typical rules are given in table 7.5.

Table 7.5 Rules to keep shaft voltages below a reasonable limit

Type of shaft voltage	Main reason for shaft voltage	Applications	Rules to minimize the voltage level	Estimated maximal influence on the shaft voltages
Circulating flux induced voltages (circulating bearing currents)	Capacitive currents within the stator winding cause a circulating flux	Converter started turbo-generator sets, medium voltage drives and larger low voltage drives	Minimize the capacity between transformer and converter, whereas the capacity between machine and converter could be higher but is limited by the traveling waves between machine and converter on voltage source converter applications	Factor 2-3 [7.10]
			Voltage drops du/dt of the converter output should be limited to 100 V/μs	
			Increase the impedance of the common mode current path by symmetrical choke arrangement in CSI, d.c. circuit inductance or by filters like common mode chokes	Factor 2-5 [7.1 7.10, 7.6]
Capacitive imposed shaft voltages	Converter supplies rotor windings	Static excitations of synchronous machines or double fed ASM	Symmetrical d.c. inverter circuit towards ground potential in order to reduce the absolute voltage peak value	[7.3]
			Limit the common voltage peaks <2 kV-2.5 kV by an adequate rated rotor winding voltage	
Capacitive imposed shaft voltages	Frequency converter feeds stator winding	Mainly smaller ASM with relatively small air-gaps	Increase the impedance of the common mode current path by symmetrical choke arrangement in CSI, d.c. circuit inductance or by filters like common mode chokes	Factor 2-5 [7.1, 7.6]
Capacitive imposed shaft voltages	Ineffective stator grounding	Converter drives in general	Equalize HF ground potential between motor frame, inverter chassis and driven machine: inverter frame-floor; motor frame –gearbox or load machine; motor frame – building steel; motor base plate – building steel; motor frame – motor terminal box	[7.6]

The rules within table 7.5 can be generally applied to converter drives under consideration of the individual needs of the plant. A turbo-generator set will not be affected by capacitive coupled shaft voltages supplied over the air-gap. The air-gap is often in a range of 50 mm-150 mm with a very small capacity. The shaft voltage due to circulating flux is caused by phenomena with time periods from 5 µs to 50 µs. Long cables between converter and generator will act as a kind of filter in case the surge is not too deep. On voltage source converter fed asynchronous machines the effect may be vice versa, [7.1]. The shaft voltages are dominated by capacitive effects over the air-gap depending strongly upon the common mode voltage peaks in the stator winding. Long cables between motor and converter may increase this effect due to traveling wave phenomena within the cable.

Chapter 8
Insulation Strategies in Converter Driven Machines

8.1 Overview of Converter Implied Insulation Stress

The repetitive switching of power electronic elements within frequency converters causes voltage peaks, which are beyond the sinusoidal voltage maximum of the fundamental. This may lead to increased electrical stress of the winding insulation. In order to avoid costly voltage filters either a special converter design or adjusted motor winding and winding insulation are often applied. These measures allow for a high reliability with long insulation life times. If the insulation system is not adjusted to the converter, damages will occur, see [8.12-8.16]. A proper winding insulation can be only designed, if the physical voltage phenomena in case of converter supplied drives are well known. Electrical motors must be protected in three main areas against electrical stresses. Within the end-winding region the phase to phase voltage U_{pp} occurs. In the active motor part the winding is placed in slots, so that the insulation must withstand the phase to ground voltage levels U_{pe}. Last but not least motors within the considered power range consist of multi-turn windings. Between two winding turns a voltage U_{turn} exists. Whereas the individual voltage levels in these three areas are clearly defined in a direct on line sinusoidal supplied motor, special effects have to be considered in case of converter supplied machines. Principally the maximum phase to phase voltage is given by the d.c. converter circuit, but this voltage level may occur between phase and ground as well. The electrical potential of the motor does not need to be symmetrical, but is determined by capacitive couplings. Another effect occurs due to switching of the electronic elements. Due to steep fronted terminal voltages shorter than 1µs capacitive effects in the winding are important, see [8.11]. The main voltage drop occurs within the first coil. If the propagation time is less than the voltage rise time a reflection takes place at the end of the first coil. This reflection causes the highest electrical stress within the inter-turn insulation to be between the last two turns of the first coil. Of course the actual voltage difference depends on the coil parameters [8.1]. In extreme cases the maximum electrical stress of the inter-turn insulation within one coil can reach up to 30% of the phase to ground voltage step [8.11]. A similar effect is known along cables between the converter and the motor. Normally the impedance of the cables is different to the motor. This will cause reflections as well. The influence of the reflections on the motor terminal voltage depends upon the cable length and the voltage rise time [8.2].

O. Drubel: *Converter Appl. & their Influence on Large Electr. Mach.*, LNEE 232, pp. 135–151.
DOI: 10.1007/978-3-642-36282-8_8   © Springer-Verlag Berlin Heidelberg 2013

8.2 Potential Distribution within a Converter Supplied Electrical Machine

The neutral point of the electrical machine and the ground potential are no longer symmetrical due to the discrete switching between the individual machine terminals with surge characteristics. The motor consists of several capacities. An overview over capacitive couplings is given in fig. 8.1.

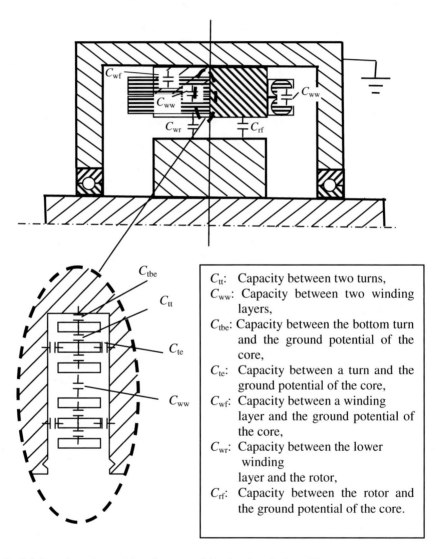

Fig. 8.1 Overview of capacitive elements within the electrical machine

8.3 Voltage Peaks due to Wave Reflection

The potential distribution within the motor is clamped by these capacities. A surge voltage, which arrives at one phase of the machine terminals, will change the potential of the phase, but it will change the voltage level of the frame in quite the same way due to the capacitive coupling. The voltage level of the phase, which is connected to the other d.c. link, gets the d.c. voltage difference as the potential difference towards the frame during this surge. Therefore the whole insulation system is stressed according the phase to phase voltage not only by the phase to ground voltage.

8.3 Voltage Peaks due to Wave Reflection

The motor, the converter and the transformer are connected by cables. All components together form a system, which consists of resistances, inductances and capacities. The system can no longer be described by just one concentrated element per component. The circuit theory has to be applied [8.9]. Fig. 8.2 shows the impedances per length within a conductor.

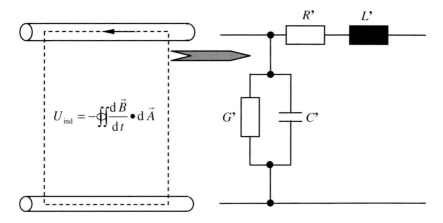

Fig. 8.2 Induced voltage and conductor impedances

The current and voltage system can be characterized by the equations 8.1, 8.2.

$$\frac{\partial^2 u}{\partial x^2} = L' \cdot C' \cdot \frac{\partial^2 u}{\partial t^2} + (C' \cdot R' + G' \cdot L') \frac{\partial u}{\partial t} + G' \cdot R' \cdot u, \tag{8.1}$$

$$\frac{\partial^2 i}{\partial x^2} = L' \cdot C' \cdot \frac{\partial^2 i}{\partial t^2} + (C' \cdot R' + G' \cdot L') \frac{\partial i}{\partial t} + G' \cdot R' \cdot i \tag{8.2}$$

Equations 8.1 and 8.2 can be simplified for quasi stationary field problems according equations 8.3 and 8.4:

$$\frac{\partial^2 u}{\partial x^2} = \left(L' \cdot C' \cdot \left(-\omega^2\right) \cdot u + \left(C' \cdot R' + G' \cdot L'\right) j\omega + G' \cdot R'\right) \cdot u \,, \tag{8.3}$$

$$\frac{\partial^2 i}{\partial x^2} = \left(L' \cdot C' \cdot \left(-\omega^2\right) \cdot i + \left(C' \cdot R' + G' \cdot L'\right) j\omega + G' \cdot R'\right) \cdot i \tag{8.4}$$

The conductor impedances change within the individual components in reality, when the cables are connected or when the coils comes out of the motor slot. Several conductor systems have to be considered, even if the system is simplified by a voltage source instead of a detailed converter and transformer model. A special impedance is defined by the lead between the terminal box and the end winding region. Afterwards the top layer of the first winding is capacitive coupled to the second turn in the end winding region. In the slot a capacitive coupling to the stator iron is given additionally. After a bend with end winding impedances, the conductor is coming back along the active part. The second turn has different conductor parameters related to the capacitive coupling and to the inductance. This means for each winding turn different impedances, which have to be determined and applied to the circuit theory. The motor winding is a multi-conductor arrangement in each slot. In case of a multi-conductor transmission line equation 8.2 becomes a matrix, where the elements of the voltage and current vectors describe the voltage and current distribution along each individual conductor, see equations 8.5 and 8.6, acc. [8.1, 8.3]:

$$\frac{\partial^2 [U]}{\partial x^2} = [Z] \cdot [Y] \cdot [U] = [P] \cdot [U] \,, \tag{8.5}$$

$$\frac{\partial^2 [I]}{\partial x^2} = [Y] \cdot [Z] \cdot [I] = [P_t] \cdot [U] \,, \tag{8.6}$$

where $[Z], [Y]$ are the impedance and admittance matrices of the line.

Equations 8.5 and 8.6 are generally solved by the following approach:

$$[U] = [U_s] \cdot e^{(-[\gamma] \cdot x)} + [U_r] \cdot e^{([\gamma] \cdot x)} \,, \tag{8.7}$$

$$[I] = [Y_0] \left([U_s] \cdot e^{(-[\gamma] \cdot x)} - [U_r] \cdot e^{([\gamma] \cdot x)}\right), \tag{8.8}$$

where $[Y_0]$ is the characteristic admittance matrix.

A modal analysis of equations 8.5 to 8.8 allows to convert the differential equations to a linear equation system. The motor system could be described by two-port network elements in a chain conductor equivalent circuit, [8.1]. The model is characterized in fig. 8.3.

8.3 Voltage Peaks due to Wave Reflection

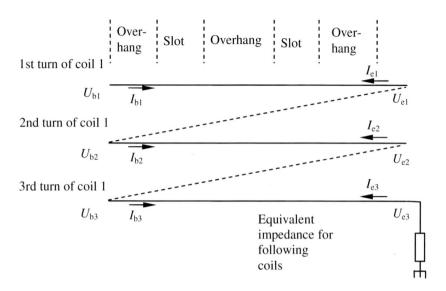

Fig. 8.3 Model of two-port networks

The model in fig. 8.4 has already been simplified. The change in the capacitive coupling to ground between the slot and the overhang in the end winding region will cause reflections as well. Only the principle system approach is shown here. Methods to determine parameters are given in [8.2].

Each individual two-port network is characterized by voltage and currents at the beginning and end of each conductor. The system of all elements is given by the following equations 8.9:

$$\begin{bmatrix} I_b \\ I_e \end{bmatrix} = \begin{bmatrix} A & -B \\ -C & D \end{bmatrix} \cdot \begin{bmatrix} U_b \\ U_e \end{bmatrix}, \qquad (8.9)$$

where

$$A = D = [Y_0] \cdot [Q] \cdot [\gamma]^{-1} \coth([\gamma] \cdot l) \cdot [Q]^{-1}$$

$$B = C = [Y_0] \cdot [Q] \cdot [\gamma]^{-1} \operatorname{cosech}([\gamma] \cdot l) \cdot [Q]^{-1},$$

l is the length of the line, $[I_b]$, $[I_e]$ are the current vectors at the beginnings and ends of the conductors, $[U_b]$, $[U_e]$ are the voltage vectors at the beginnings and ends of the conductors, $[Q]$, $[\gamma^2]$ are the Eigen-vectors and Eigen-values of $[P]$.

Equation 8.9 has twice the number of unknowns than equations. The following relationships according fig. 8.3 have to be considered as well:

$$I_{e1} = -I_{b2}; \quad I_{e2} = -I_{b3}; \quad I_{e3} = -U_{e3}/Z; \quad U_{e1} = U_{b2}; \quad U_{e2} = U_{b3}.$$

140   8 Insulation Strategies in Converter Driven Machines

The whole approach allows the calculation of the voltage at the motor terminals and at the turn connections within an electrical machine. The voltage of the converter is Fourier transformed for this approach and applied to the systems of two- port networks as first network voltage [8.1]. The impedances do depend on the frequency.

Even though the complete description of the voltage distribution within converter-motor systems requires a solution of the two-port network, some basics for the voltage at the motor terminals can be understood by investigations to a homogenous conductor, which is open ended. The wave reflection at the change in impedances between the cable and the motor winding is often defined by a factor of two. This factor is derived by the solution of equation 8.2 at the motor terminals for a homogenous conductor. The relationship of the voltage amplitude at the motor terminals U_2 and the amplitude of the positive wave U_s are given by equation 8.10:

$$U_2 = U_s \cdot \left(1 + \frac{Z - Z_0}{Z + Z_0}\right),$$
(8.10)

where Z_0 is the cable impedance and Z is the motor impedance at the terminals.

The factor two is derived, if $Z \gg Z_0$. It is very important, that equation 8.10 is only valid, if the voltage U_1 at the converter terminals consists only of the positive wave U_s. The positive wave is not always the same as the voltage at the converter. Therefore the voltage at the motor terminals can be more than a factor two higher than the converter voltage. The voltage distribution of an open ended conductor is described in equation 8.11, [8.10]:

$$u(x) = U_s \cdot \exp(-j\beta(x-l)) + U_r \cdot \exp(j\beta(x-l))$$
$$= \underbrace{U_s \cdot (\exp(-j\beta(x-l)) + \exp(j\beta(x-l)))}_{\text{open ended conductor}} = U_2 \cos(\beta(l-x)),$$
(8.11)

with $\beta = \mathrm{Im}\left\{\left(\sqrt{(R'+j\omega L')(G'+j\omega C')}\right)\right\} \approx \omega\sqrt{L'C'}$.

The converter is placed at the position: $x = 0$. The relation between the motor and converter terminal voltages is given in equation 8.12:

$$\frac{U_2}{U_1} = \frac{1}{\cos(\beta l)}.$$
(8.12)

Long cables together with short voltage rise times will lead to voltage peaks, which can be more than two times larger than the d.c. link voltage. The amplitude at the motor terminals will stress especially the first winding coil. The transient potential distribution may stress the interturn insulation between the wires or strands of the first coil. Longer cable length will increase the terminal voltage

8.4 Influence of the Voltage Characteristic on the Insulation Life Time

peaks, but the surge time will decrease. The stress will depend on the machine parameter, but values in the range of 50 % or more of the motor terminal voltages have to be expected for surge times of 100 ns. Smaller values of around 20 % may be expected for surge times in the range of 1 μs.

8.4 Influence of the Voltage Characteristic on the Insulation Life Time

The effect of the converter caused electrical stresses within the motor itself depends upon the type of insulation. Motors, which are built for medium-voltage applications, are rated for the system voltages to which they are applied. They may have some extra insulation thickness for converter operation. The winding consists typically of form wound coils. Form wound coil windings have insulated rectangular strands. They are carefully taped or wrapped. The presence of air between the turns is minimized due to the regular layers. Vacuum pressure impregnated (VPI) processes allow for substitution of the remaining small areas with air by varnish. Partial discharge between the turns is kept to a minimum, because it can not occur without the presence of air. Ionization takes place in air due to the fact that the breakdown field voltage in air is lower than in the insulation. The voltage level between two adjacent turns is defined by the turn voltage and not by the phase voltage. The regular winding of the turns defines the positions of the turns definitely. They are in sequential order and turn 1 touches turn 2 and turn 2 will only touch turn 1 and 3 and so on.

One source for partial discharge in form wound coils does still exist. Thermalmechanical aging effects may allow for small spaces between the coil and the iron lamination of the stator core. This space will be filled with air, so that partial discharges could occur between the iron surface and the outer insulation layer.

Low voltage systems are realized by different technologies. Windings consist of random wound coils. These coils are wound on a special tooling and inserted in the stator slots afterwards. It may occurs, that the first and last turn of the coil are closely together or even touch due to the manufacturing process. Adjacent turn voltage would be full coil voltage in this case. If the end turns of adjacent coils within random wound windings are not separated by special insulation sheets, even higher voltage levels up to the phase voltage could determine the maximum voltage level over the turn insulation in the end winding region. Even though random wound low voltage systems are often vacuum pressure impregnated, the numerous crossovers combined with the round wire geometry makes it nearly impossible to eliminate all air pockets in the coils. The little air pockets allow for partial discharges within the coils [8.4], if the field strength is over the breakdown strength.

Interturn insulations of low voltage windings are more at risk than those of medium voltage windings due to the different technology processes. Therefore the influence of converter caused voltage pulses on the insulation will be shown on a low voltage system. Specialities, which may cause damages to medium voltage systems, will be handled afterwards.

The influence of converter caused pulse voltages on low voltage winding insulation systems is investigated excellently in great detail in [8.5]. In a first step breakdown mechanisms are determined in case of a pulse voltage. An overview of the number of surges to failure as function of the peak uni-polar impulse voltage is given in fig. 8.4. Two regions can be identified. The curve shows a linear characteristic in double logarithmic presentation in the region of voltages above 2 kV. The curve bends for voltages below 2 kV and reaches pulse rates above 10^{10} for voltages below 1.4 kV. The breakdown mechanism is based in both regions on partial discharge processes. The number of pulses until a breakdown occurs depends on the growing speed of the material distortions for voltages above 2 kV. It depends upon the probability of first partial discharges for lower voltages. The number of surges until a failure occurs is influenced by the pulse form. A voltage pulse can be described according to fig. 8.5.

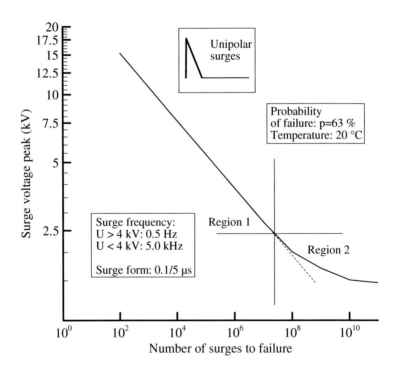

Fig. 8.4 Number of surges to failure as function of peak voltage acc. [8.5]

8.4 Influence of the Voltage Characteristic on the Insulation Life Time

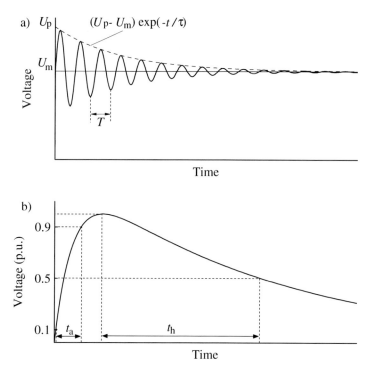

Fig. 8.5 Definition of oscillating pulses, surge rise time t_a and pulse half time t_h, acc. [8.5]

The electrical field strength in air pockets is influenced by charging processes at the insulation surfaces. Two main kinds of charge generation can be found. In case of a partial discharge electrons are generated, which are traveling to the anode. Indeed they will not reach the anode, but end up on the insulation varnish. The second kind of charge generation does not depend on a partial discharge. Charging processes occur as well due to interactions at surfaces with limited speed. The charges are either from the surrounding air, of the varnish or from rigid bodies, which release their charges due to the high field strength. The charge agglomeration will cause in a first step a reduction of the field strength in air. This reduction will not occur to the same extent, if the voltage surge time t_a is very short. The shorter the surge time is, the less the possible reduction in field strength will be and the higher is the probability for a partial discharge. The influence of the surge time and pulse half time on the time to failure is given in fig. 8.6.

The pulse half time has only influence in the area, where not 100 % of all surges pulses cause partial discharge. This is in the case <2 kV. The longer the pulse half time is, the longer the insulation will be stressed by the relatively high

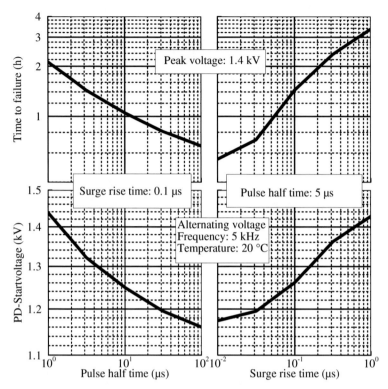

Fig. 8.6 Time to failure and partial discharge start voltage as function of pulse half time and surge rise time, acc. [8.5]

field strength and the higher is the probability for a partial discharge. The time to failure reduces by a factor of three in case of a longer pulse half time of a factor ten, see [8.5].

This phenomenon is superposed in case of an alternating voltage. Whereas the charge agglomeration reduces the field strength during the first pulse, the charges will increase the field strength during the pulse with opposite polarity, see fig. 8.7.

The increase in field strength due to the alternating voltage characteristic has been shown in fig. 8.7. Quantitative effects of the phenomenon are given in fig. 8.8.

The characteristic curves in fig. 8.8 are shifted due to the direct influence of the increase in field strength with the alternating voltage surge.

8.4 Influence of the Voltage Characteristic on the Insulation Life Time

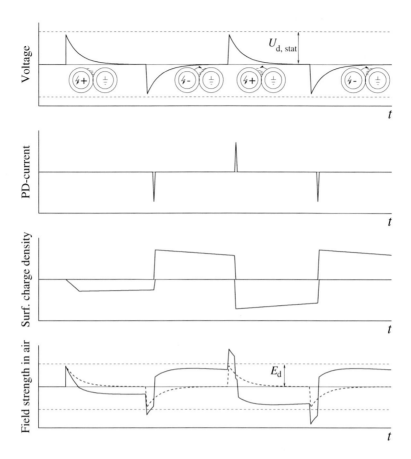

Fig. 8.7 Charging processes, partial discharge and field strength for alternating voltages between turns, acc. [8.5]

Beside the influence of the pulse form and the difference between alternating and uni-polar pulses the stress in the insulation during a surge and during a sinusoidal supply should be compared. A comparison is given in fig. 8.9, which is based on the peak values of the pulse and the amplitude of the sinusoidal voltage. The pulse repetition frequency is given by f and complies with the time distance between two half waves of a sinusoidal supply. The number of pulses until a failure occurs is about a hundred times higher for sinusoidal voltage forms than for pulse forms with voltage levels above 2 kV. Both characteristics show a similar characteristic for voltages below 1 kV. Voltage levels below 1 kV are necessary for the test coils in order to reach times to failure above 1000 h. Even though the

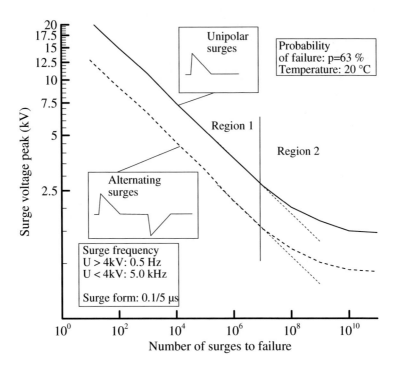

Fig. 8.8 Number of surges to failure for uni-polar and alternating voltage surges, acc. [8.5]

areas below 1000 h may be of interest for scientific purposes, the area above 1000 h is to be considered for the practical applications of electrical machines.

Due to these results it is important to dimension the insulation system in a way, which allows to reach the times to failure, where the breakdown mechanism do no longer depend on the surge rise time.

A summary of the results may be given as follows:

- Sinusoidal and pulse voltages cause a breakdown after similar numbers of pulses or half waves in case that the voltage is so high, that a partial discharge occurs with any half wave.
- The plant design has a strong influence on the voltage step at the motor terminals and determines the maximum allowable converter d.c. link voltage for an existing motor insulation system, see [8.17]
- The time until breakdown is similar for sinusoidal and pulse voltages in the case that the voltage is low enough, that hardly any partial discharge occurs.

8.4 Influence of the Voltage Characteristic on the Insulation Life Time 147

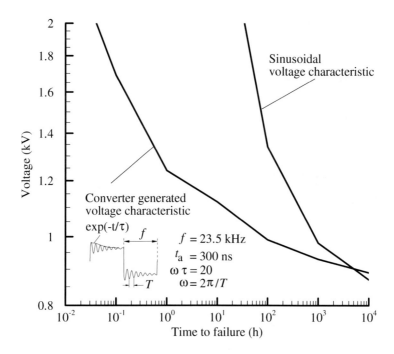

Fig. 8.9 Comparison of the time to failure for a sinusoidal voltage characteristic and a comparable converter generated voltage, acc. [8.5]; the pulse repetition frequency f complies with twice the frequency of the sinusoidal voltage signal

Whereas most of the found rules are valid for form wound coils as well, some more effects have to be considered especially in the end winding region. Fig. 8.10 gives an impression about the principle insulation system in this area. The individual medium voltage coils are soldered at their end turn. The surface at these places reaches nearly the potential of the copper whereas the potential in the active part is defined by the laminated iron. The potential difference is controlled over a stress grading coating in order to avoid creeping currents.

A typical distribution of the electrical field along the surface of the stress grading coating is given in fig. 8.11.

The potential control with a stress grading coating in comparison with a constant low conductivity is shown. This insulation system works very well for sinusoidal voltages. The situation is different for pulse voltages. A transient voltage surge will cause a concentration of field strength at the slot-outlet in the area of the semi-conductive layer, see fig. 8.12. The conductivity of the semi conductive layer could be increased in order to reduce the high field strength at the slot-outlet. Unfortunately the conductivity of the semi-conductive layer is limited in the active part due to alternating leakage flux, which would cause eddy currents and overheating. Additionally the lamination would be short circuited, which would mean eddy currents in the lamination as well. A solution, which can be applied, is to

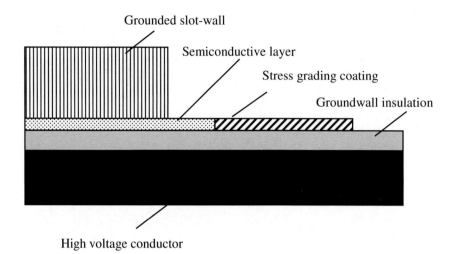

Fig. 8.10 Principle insulation system for the end winding region in form wound medium voltage machines, acc. [8.6]

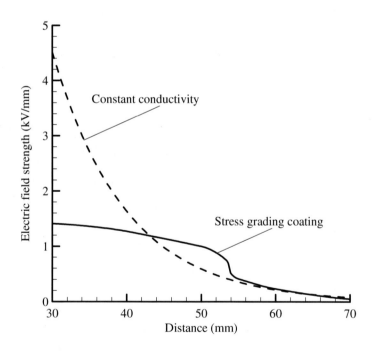

Fig. 8.11 Field strength along a stress grading coating and a coating with constant conductivity, with 10^{-6} S/m, for a 60 Hz 24 kV sinusoidal voltage, acc. [8.6]

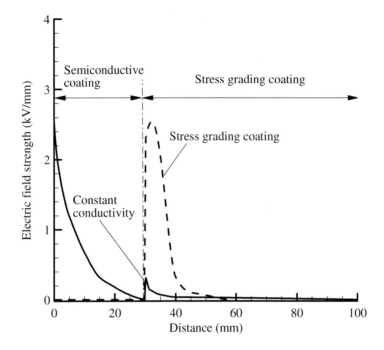

Fig. 8.12 Electric field from the slot-outlet to the end of the stress grading coating at 200 ns for a semi-conductive coating of 0.02 S/m and 1 S/m, acc. [8.6]

increase the conductivity only in the area outside the slot. Fig. 8.12 shows the reduction of the maximum field strength from 2.5 kV/mm by about 10 % to 2.2 kV/mm if the conductivity at the semi-conductive area is increased from 0.01 S/m up to 1 S/m.

8.5 Influence of the Converter Type on the Insulation System

Frequency converter imposed voltage characteristics cause additional electrical stress to the insulation system due to four main reasons:
- Traveling waves on the cable system or within the winding of the electrical machine can cause high voltage peaks
- High frequency voltage oscillations occur on the drive-motor system which increase the number of voltage peaks per electrical switching process. Some of the oscillations cause several alternations of the voltage polarity within a half wave of the fundamental frequency.
- Less relatively slow charging processes, which reduce the electrical field strength in the small air holes between conductors, will take place due to too short surge rise times. This effect is only important, if the absolute field strength is in a range, where partial discharge occurs.
- The insulation between the winding and ground is exposed to higher electrical stress levels due to capacitive couplings and the transient field distribution.

Fig. 8.13 gives an impression of surge characteristics for three different converter types. The voltage has been measured at the motor terminals to ground. The actual rated voltages of the converter types differ. The two-level converter is a low voltage converter, the three-level and multi-level converters are medium voltage converters. In order to get a proper comparison the measured voltage characteristics have been divided by the amplitude of the fundamental harmonic of each measured characteristic. Motor insulation life times are between 160000 h and more than 700000 h for industrial applications. Shorter time periods can not be accepted because motors are designed for a life time of more than 20 years. In order to achieve these figures the absolute insulation stress has to be small enough to avoid damages due to partial discharge. Damages due to partial discharge will only occur after the insulation becomes weaker due to ageing effects. The surge pulse time and the pulse half time hardly influence the life time of such a proper designed insulation system. The influence of the capacitive couplings to ground is taken into account by a lower conductivity in the range of the slot-outlet and by an insulation system, which can endure phase to phase voltages between the winding and ground in the slots. This is especially important for the two- and three-level converter with short surge rise times. The surge rise time and the surge amplitude of the multi-level converter is such, that no special measures would be necessary

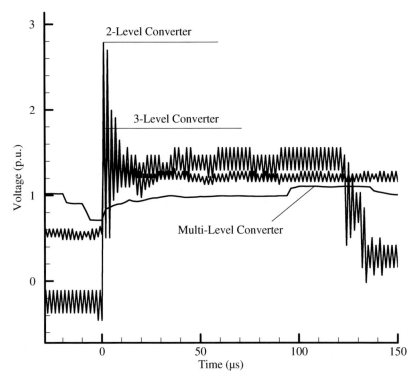

Fig. 8.13 Comparison of three different voltage converter designs

8.5 Influence of the Converter Type on the Insulation System

to the insulation system. Beside the surge rise time the amount of pulses, which have to be considered for the insulation stress, is important as well. The two-level converter causes an alternating voltage with each pulse and voltage oscillation on the system, which are not alternating. In case of the cascaded three-level converter the zero voltage level is used. The voltage characteristic is uni-polar during a half period. The alternating polarization has to be taken into account for the two-level converter with the pulse frequency. It has to be considered for the three-level and multi-level converter for the fundamental frequency.

Last but not least the peak voltage of all three converters related to the amplitude of the fundamental is different. This is the most important difference, because the maximum voltage will determine, whether partial discharge occurs at all. Whereas the multi-level converter causes peaks of only about 1.1 times of the fundamental voltage amplitude, peaks of 1.8 times and 2.8 times of the fundamental voltage amplitude have been measured for the three- and two-level converters. Insulation systems of motors, which are operated on the three- and two-level voltage source converters, must be designed to cope with these peaks.

Chapter 9
Converter Applications in Typical Examples of Energy Efficient Pump Systems and in Processes with a Strong Overload Characteristic

9.1 Energy Saving Potential for Boiler Feed Water Pumps in Part Load Operated Coal Power Plants for Different Drive Arrangements

Converter driven electrical machines are mainly used for pumps and compressor loads or applications with special overload requirements and control quality. Converter driven machines for pumps can be found for instance in several electrical steam or combined cycle power plants. Combined cycle power plants and steam power plants are more and more in focus for an acceptable load control within energy grids. Load profiles and plant control strategies are optimized taking the market liberalization and the strong contribution of renewable sources into account. Uncertainties in the power supply in grids with a large renewable wind energy component will request an adequate amount of power plants, which act as reserve. Different kinds of reserve power need to be supplied in an electrical grid. A primary reserve needs to be activated within a time of 20 s. A secondary reserve must supply the grid after 5 min with electrical energy and a minute reserve needs to be available in order to supply power for a longer time period, when only minor wind sources are available. Especially the primary and secondary reserves need special plants or operation schema. A first control instant is given by water storage plants, [9.1]. Storage water plants have been built for the primary and secondary control requirements of an electrical grid at different places, but pump storage can contribute only to a very small amount as compensation to the wind variations. Special countries like Switzerland or Austria have enough water storage power plants due to the Alps, but countries like Germany f.i. have only 3 % of their installed plant capacity in pump storage plants. Variations in wind power of 5-15 GW within a quarter of an hour require much more secondary and minute reserves. Combined cycle plants are a solution for this request. Combined cycle power plants need gas or oil for their burners. Unfortunately these fossils are

O. Drubel: *Converter Appl. & their Influence on Large Electr. Mach.*, LNEE 232, pp. 153–160.
DOI: 10.1007/978-3-642-36282-8_9   © Springer-Verlag Berlin Heidelberg 2013

world wide limited to some few countries outside of Europe and reserves, which can be gained with reasonable efforts, will be exhausted within the next decades. Ongoing developments may allow in future a proper gasification of coal. The gas will be burned in gas-turbines afterwards. A generation of hydrogen out of water during periods with large wind energy contribution and the supply to gas-turbines of this hydrogen, when the wind contribution is less, may evolve in future, [9.2]. The losses during the conversion would be used for local heating. So far the overall process is neither efficient enough nor cost effective. The energy reserve of storage water power plants and combined cycle plants is often not sufficient within the German electricity grid. Conventional steam power plants are used as reserve as well. A plant, which needs to supply secondary control power, must be operated in part load. Coal fired power plants need much more than 5 minutes for a cold start. A steam power plant can operate to the secondary control, if the plant is operated in part load before. A typical coal power plant is schematically shown in fig. 9.1.

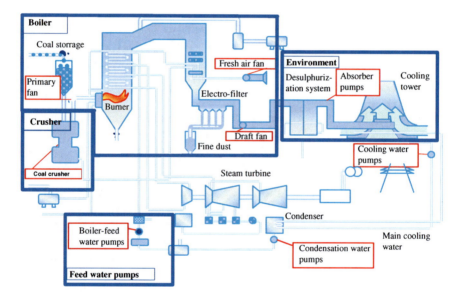

Fig. 9.1 Typical configuration of a coal fired steam power plant

Several large electrical motors are installed for auxiliary processes beside the main turbine group with the turbo-generators. About 7% of the rated plant power is needed to feed these processes. Main energy consumption occurs with the boiler feed water pump with about 4 %. In case of power plant part load operation the volume flow through these pumps must be reduced. The thermodynamic process is determined by the volume flow, the water pressure and the temperature, which will be reached in the boiler. The temperature determines the efficiency. Pressure and volume flow influence mainly the amount of energy generation. The boiler is fed with a reduced amount of coal in part load operation. The reduced heat flow

9.1 Energy Saving Potential for Boiler Feed Water Pumps

would reduce the process temperature and the efficiency, if the volume flow is kept constant. The volume flow must be adjusted in the same way as the heat flow, which can be achieved by different arrangements, see fig. 9.2. Indeed each arrangement has its advantages in efficiencies and costs, which depend upon the plant operation. Two basic principles can be applied to reduce the flow. Either the pump characteristic is kept constant and the operation point is modified by some orifices in case a) or the pump characteristic is modified by a speed variation of the pump, case b)-e).

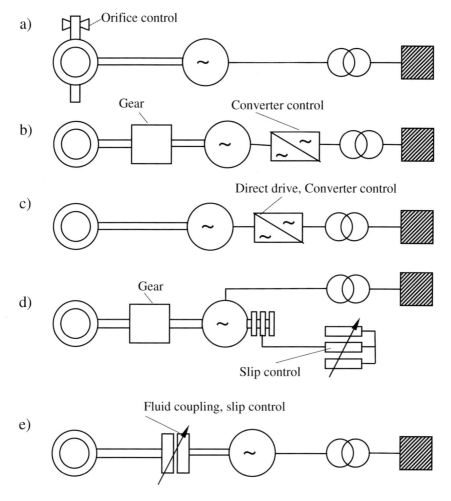

Fig. 9.2 Plant arrangements of boiler feed water pumps; a) with orifice control; b) frequency converter controlled cage induction motor with gear box; c) direct drive frequency controlled cage induction motor; d) with slip ring induction motor and gear box; e) with fluid coupling

156 9 Converter Applications in Typical Examples

The required pump power will change according to equations 9.1 and 9.2 with the volume flow:

$$\text{Case a): } P_{pump} \propto \dot{V},\tag{9.1}$$

$$\text{Case b), c), d), e): } P_{pump} \propto \dot{V}^3 \propto n^3.\tag{9.2}$$

The difference between equation 9.1 and equation 9.2 will be equivalent to the losses within the orifices, see equation 9.3:

$$P_{loss} = P_{pump, \ r} \cdot \left(\frac{\dot{V}}{\dot{V}_r} - \left(\frac{\dot{V}}{\dot{V}_r} \right)^3 \right).\tag{9.3}$$

The required motor power in case d) and e) will increase in comparison with the required pump power with the slip according to equation 9.4:

$$\text{Case d), e): } \left(1 - s\right)P_\delta = P_{pump}.\tag{9.4}$$

The additional losses can be calculated according equation 9.5:

$$
\begin{aligned}
P_{loss} = P_\delta - P_{pump} &= \left(\frac{s}{(1-s)} \right) P_{pump} \\
&= \left(\frac{s}{(1-s)} \right) \left(\frac{\dot{V}}{\dot{V}_r} \right)^3 P_{pump,r} = \left(\frac{\dot{V}_r}{\dot{V}} - 1 \right) \left(\frac{\dot{V}}{\dot{V}_r} \right)^3 P_{pump,r} = sP_\delta
\end{aligned}
\tag{9.5}
$$

Beside the plant arrangement itself, the converter type and motor design will influence the amount of losses. Both motor and converter build a system, which influence each other. Converters with high pulse frequencies and multiple voltage levels are capable to a nearly sinusoidal voltage characteristic. Motor losses are similar to sinusoidal supply due to such a characteristic. The converter losses itself do depend upon the converter topology and pulse frequency. The higher the pulse frequency, the higher are the converter losses. The following fig. 9.3 gives an overview of converter-, motor efficiency and the efficiency of the system converter-motor for different motor loads for a nominal pump power in the range between 2 MW-10 MW. The converter losses consist of a component, which depends on the current amplitude and a component, which depends on the frequency of switching. The current depending losses will be reduced for a power reduction at constant speed. The loss reduction depends on the non-linear voltage-current characteristic of the semi-conductive elements. Though the losses will be reduced, the efficiency is nearly constant, because the load is reduced as well. The efficiency of the motor will be quite constant up to 60 % of the rated load. The friction- and no-load losses become more and more dominant for further load reductions. Even though main loss components are given according equations 9.1-9.5 and the losses within the motor-converter system, losses will occur in the gear box as well. Gear box efficiencies are normally in the range of converter efficiencies. Table 9.1 gives an overview of the different amount of losses for the cases a)-e).

9.1 Energy Saving Potential for Boiler Feed Water Pumps

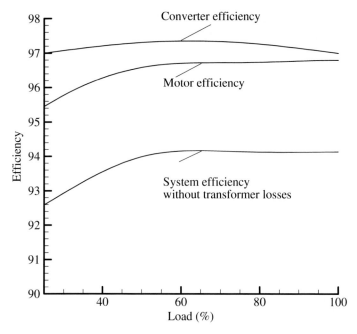

Fig. 9.3 Converter, motor and system efficiency versus motor load

Table 9.1 Overview of losses for different pump arrangements for part load and rated load

Case	Equation	Loss related to rated pump power for 70% rated volume flow	Loss related to rated pump power for rated volume flow
a)	$P_{loss} = \left(\left(\dfrac{\dot{V}}{\dot{V}_r} - \left(\dfrac{\dot{V}}{\dot{V}_r}\right)^3 \right) + \dfrac{(1-\eta_{motor})}{\eta_{motor}} \dfrac{\dot{V}}{\dot{V}_r} \right) \cdot P_{pump,r}$	40%	5%
b)	$P_{loss} = \dfrac{(1-\eta_{motor}\eta_{gear}\eta_{conv})}{\eta_{motor}\eta_{gear}\eta_{conv}} \left(\dfrac{\dot{V}}{\dot{V}_r}\right)^3 \cdot P_{pump,r}$	6%	11%
c)	$P_{loss} = \dfrac{(1-\eta_{motor}\eta_{conv})}{\eta_{motor}\eta_{conv}} \left(\dfrac{\dot{V}}{\dot{V}_r}\right)^3 \cdot P_{pump,r}$	3%	7%
d)	$P_{loss} = \left(\left(\dfrac{\dot{V}_r}{\dot{V}} - 1\right)\left(\dfrac{\dot{V}}{\dot{V}_r}\right)^3 + \dfrac{(1-\eta_{motor}\eta_{gear})}{\eta_{motor}\eta_{gear}} \right) \cdot P_{pump,r}$	22%	8%
e)	$P_{loss} = \left(\left(\dfrac{\dot{V}_r}{\dot{V}} - 1\right)\left(\dfrac{\dot{V}}{\dot{V}_r}\right)^3 + \dfrac{(1-\eta_{motor})}{\eta_{motor}} \right) \cdot P_{pump,r}$	19%	4%

9.2 Influence of Speed and Overload Capability of Steel Mill Drives on the Motor Efficiency

Pump and compressor applications are always dimensioned to their performance at the maximum speed, due to the fact, that the required motor power is proportional to the cube of the speed. Other applications require a constant motor power or motor torque independent on the speed. A typical load characteristic for a steel mill application is given in fig. 9.4.

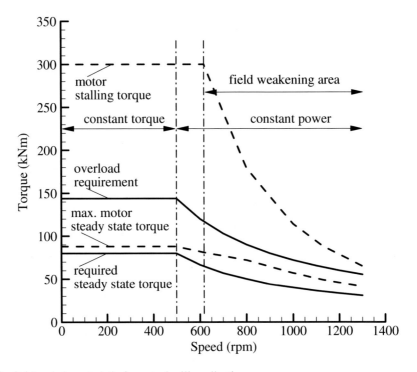

Fig. 9.4 Load characteristic for a steel mill application

The load characteristic requires an optimization of the converter motor system. The converter is dimensioned by the base frequency of the motor. The base frequency defines the frequency, where the nominal terminal voltage is reached. A converter optimum is reached, if the base frequency coincides with the transition between the constant torque and constant power requirement.

The motor stalling torque is defined by its overload capability within the field weakening area. If the field weakening area due to the converter choice is wide together with the need for overload capabilities, the stalling torque may require a relatively large motor size. The overload capability is in several processes only used for a short time period and the motor would be operated most of the time

9.2 Influence of Speed and Overload Capability of Steel Mill Drives

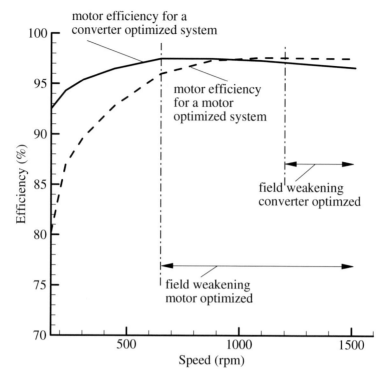

Fig. 9.5 Motor efficiencies dependent on speed for a system, which is optimized to the converter size, in comparison with a system, which is optimized to the motor size

in part load. The motor can be optimized, if the base frequency is placed in the constant power area according to the motor's thermal limitations. Fig. 9.5 shows the motor efficiency in line with the load characteristic for an application, where the converter has been optimized in size and for the case, where the motor has been thermally optimized in size.

If the motor is optimized in size, the losses in the process are differently distributed. This is shown in fig. 9.5 as well. Obviously it is better to optimize the converter, than the motor. The motor efficiency is only in the speed range above 1000 rpm slightly better than for the converter optimized solution. The motor optimized solution would cause a maximum stator current more than twice larger than for the converter optimized system. The converter losses would increase in the constant torque area by a factor of two accordingly. A better approach would be to improve the system efficiency of the motor in the converter optimized system. Better iron quality may contribute by 0.2-0.3 % in efficiency within the field weakening area. The system efficiency could often be improved further, if the motor cooling would be adjusted to the thermal requirements. Due to the constant torque area external ventilation should be foreseen. This ventilation could be controlled with the motor temperature and adjusted in speed accordingly.

The number of poles does influence the efficiency as well. An overview of the motor efficiencies as function of the rotor surface speed is given in fig. 9.6.

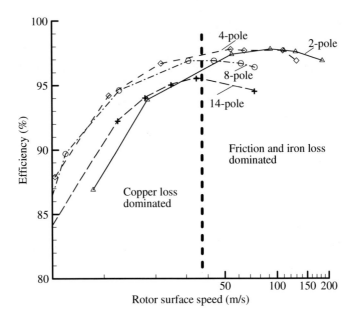

Fig. 9.6 Efficiency of 2-, 4-, 8-, 12-pole machines as function of the speed at the rotor surface

The bore diameters of the 2-pole, 4-pole and 8-pole machines are different in size, whereas the bore diameter of the 14-pole machine is the same as the one of the 8-pole machine. Low surface speeds mean a reduction of the induced voltage. Copper losses dominate the overall loss characteristic. The 2-pole machine has longer end winding regions and less space for copper in the active part due to long distances of the magnetic flux distribution than the 4- or 8-pole machines. If the speed is higher, iron losses and friction losses become more dominant. Low frequencies in the stator will reduce the iron losses. In principle a 4-pole converter driven machine is often the best choice.

Chapter 10
Conclusion

Applications of converters in the power range between 500 kVA and 200 MVA are used in several industry segments. Converter driven electrical machines are often used for pumps and compressor loads or applications with special overload requirements and control quality. The special voltage waveform and the variable frequency impose special design measures on large electrical machines. Large electrical machines implement different constraints on converter applications than smaller machines. Low voltage applications have f.i. cost advantages for the power electronics, but large machines with rotor surface velocities of 100-300 m/s require already in the power range between 1500-2000 kW medium voltage supplies. Several topologies are explained, which are industrially applied in this power range. Whereas two-level voltage source converters dominate the low voltage range, three- and multi-level converters are used in the medium voltage range. Naturally commutating converters are applied to allow for better efficiencies and lower costs per unit power. Voltage waveforms of different realized converter topologies are shown and analysed. The variability of the motor supply voltage is enormous. Calculation methods are available to consider the converter specific voltage waveform. The calculation methods allow for different levels of accuracy. Whereas naturally commutated converters like diode rectifiers in large turbogenerators are clearly defined by the power electronic elements, forced commutated converters are determined by the converter control. Calculated and measured currents agree well in naturally commutated converters, if numerical methods are applied. Forced commutated converters are in most of the cases operated with precalculated pulse cycles. The detailed control depends on the frequency and voltage of the operation point. Necessary data for a system model are hardly available, if the converter and the motor are not from the same supplier. A measured voltage of a single operation point is often the best available data. The influence of the individual converter can be estimated based on a Fourier analysis of the measured or calculated voltage waveform at the relevant operation points.

Additional losses can be determined in laminated and massive iron parts as well as in the stator winding and rotor copper in good agreement by a mixture of analytical methods and experience or simplified numerical models. Complete numerical based system calculations are not necessary in most of the cases.

Converter caused torque oscillations do require a detailed calculation of the air-gap torque harmonics and the shaft oscillations. The magnetic torque calculation can be either done analytically or numerically. The harmonic content of the torque depends strongly on the converter type. Multi-level converters allow for a quality of the voltage waveform, which limits converter-imposed torque oscillations below the amplitudes of synchronous torques at direct online operation. Two-level converter designs may create harmonic torque content with more than 10 % of the rated torque.

The prediction of the noise based on electromagnetic sources in case of converter operation is a challenge. The calculation process consists of a calculation of the motor terminal voltage, the solution of the machine field problem with a force density distribution in the air-gap and a structure mechanical machine analysis together with its noise transmission through the air. Specialties of real motors are revealed. Motors, which are clamped to a foundation for instance, will become asymmetric from a mechanical point of view. A third mechanical mode may be excited by a second or fifth force mode. This will not occur if a cylindrical shell is assumed. A converter driven motor will be excited in any Eigen-frequency, which exists between 500 Hz – 2000 Hz in most of the cases. Special rules are given, which help to reduce the vibration amplitudes nevertheless.

Shaft voltages in converter driven large machines are mainly induced by circumferential flux. This kind of shaft voltage can reach values of more than 200 V_{pp} in large turbo-generators during the gas turbine start. The shaft voltage is induced, with the capacitive stator coupling as primary- and the shaft as secondary-winding. Converters are often directly connected to the rotor of double fed asynchronous machines in wind turbine applications. The shaft voltages, which occur in this case, are directly capacitive imposed.

The repetitive switching of power electronic elements within frequency converters causes voltage peaks, which are beyond the sinusoidal voltage maximum of the fundamental. This may lead to increased electrical stress of the winding insulation. The electrical potential of the motor does not need to be symmetrical, but is determined by capacitive couplings. Another effect occurs due to switching of the electronic elements. Within the end-winding region an increase in the phase to phase voltage occurs. In the active motor part the winding is placed in slots. The insulation must withstand the phase to ground voltage levels with sinusoidal supply, but the phase to phase voltage in case of converter supply to cope with the capacitive effects. Last but not least motors within the considered power range consist of multi turn windings. Between two winding turns a voltage exists, which may be up to 30 % higher due to the voltage wave reflections during converter operation.

The different converter imposed effects within the electrical machine influence the right machine design and choice. Additionally the electrical machine has to be adjusted for the individual application. Two dominant types of applications are pump or compressor drives and constant load applications like in steel mills. Best efficiencies are reached for pump applications at nominal speed for direct online drives until the pump has to be operated in part load. In pump applications, which

10 Conclusion

do not exactly operate at nominal speed with the nominal pressure and flow, converter operated drives are best in efficiency. It may be necessary to operate the pump not only due to the part load requirements at a different operation point, but also due to deviations of the originally calculated pressure drop of the flow consumer. Constant load applications like steel mills need converter driven motors mainly due to the control quality. Nevertheless it is possible to optimize the system efficiency of transformer, motor and converter. Additionally the motor itself can be optimized in efficiency by an appropriate choice of the number of poles.

Attachment

A.1 Parameters of Asynchronous Machines

Main asynchronous parameters are given by the stator quantities as resistance R_1, the main inductivity L_1, the slot leakage L_{N1}, the end winding leakage L_{S1}, the mutual inductivity $M_{2,1}$ and the rotor quantities as the resistance in the ring R_{R2}, the resistance in the rotor bar R_{S2} the main inductivity L_2 the factor for leakage within the air gap σ_{d2}, the slot leakage L_{N2}, the end ring leakage L_{R2} and the mutual inductivity ${}^P M_{1,2}$. Equations A.1-A.10 allow to determine the individual resistances and inductances of the complete machine.

$$R_1 = w \cdot \frac{2(l_{act} + l_{S1})}{a^2 \gamma_1 A_{cu,1}}, \tag{A.1}$$

$$R_{S2} = \frac{l_{act}}{\gamma_2 A_{cu,2}} \cdot k_r, \tag{A.2}$$

$$R_{R2} = \frac{l_{R2}}{\gamma_2 A_{cu,R2}}, \tag{A.3}$$

$${}^P M_{2,1} = \frac{w}{a} l_{Fe} r_i \frac{\mu_0 Z_2}{\delta_i \cdot 2\pi \cdot p^3} {}^P k_w \cdot 2 \sin\left(p \frac{\pi}{Z_2}\right), \tag{A.4}$$

$${}^P M_{1,2} = {}^P M_{2,1} \frac{2pm}{Z_2}, \tag{A.5}$$

$${}^P k_w = \frac{\sin\left(\dfrac{\pi}{2m}\right)}{q \sin\left(\dfrac{\pi}{2mq}\right)} \cdot \sin\left(\frac{\pi}{2} \cdot \frac{y}{\tau_p}\right), \tag{A.6}$$

$$^{P}L_2 = l_{Fe} r_i \frac{\mu_0 Z_2}{\delta_i \cdot 2\pi \cdot p^2} \cdot \left(2\sin\left(p\frac{\pi}{Z_2}\right)\right)^2, \tag{A.7}$$

$$^{P}L_{N2} = \mu_0 l_{Fe} \, ^{P}\lambda_{N2}, \tag{A.8}$$

$$\lambda_{N2} = \frac{h_S}{b_S} + \frac{h_n}{3b_n}, \tag{A.9}$$

$$^{P}L_{R2} = \mu_0 \cdot 0.37 \cdot 2\pi \frac{r_r}{Z_2}, \tag{A.10}$$

$$^{P}L_1 = l_{Fe} r_i \frac{\mu_0 2}{\delta_i \cdot \pi} m w^2 \cdot \frac{^{P}k_w^2}{a^2 p^2}, \tag{A.11}$$

$$^{P}L_{S1} = 2\mu_0 l_S \frac{w^2}{a^2} \lambda_{S1}, \tag{A.12}$$

$$\lambda_{S1} \approx 0.30, \tag{A.13}$$

$$^{P}L_{N1} = 2\mu_0 l_{Fe} \frac{w^2}{a^2 \cdot p \cdot q} \cdot \lambda_{N1}, \tag{A.15}$$

$$\lambda_{N1} = k_1 \frac{h_{11}}{3b_{1n}} + k_2 \frac{h_{12}}{b_{1n}} + \frac{d_1}{4b_{1n}}, \tag{A.16}$$

$$k_1 = \left(\frac{9}{16}\frac{y}{\tau_p} + \frac{7}{16}\right) \quad \frac{2}{3} \le \frac{y}{\tau_p} \le 1.0, 60° \text{ zones}, 3 - \text{phases}, \tag{A.17}$$

$$k_1 = \left(\frac{18}{16}\frac{y}{\tau_p} + \frac{1}{16}\right) \quad \frac{1}{3} \le \frac{y}{\tau_p} \le \frac{1}{3}, 60° \text{ zones}, 3 - \text{phases}, \tag{A.18}$$

$$k_1 = \left(\frac{9}{16}\frac{y}{\tau_p} + \frac{4}{16}\right) \quad 0 \le \frac{y}{\tau_p} \le \frac{1}{3}, 60° \text{ zones}, 3 - \text{phases}, \tag{A.19}$$

$$k_2 = \left(\frac{12}{16}\frac{y}{\tau_p} + \frac{4}{16}\right) \quad \frac{2}{3} \le \frac{y}{\tau_p} \le 1.0, 60° \text{ zones}, 3 - \text{phases}, \tag{A.20}$$

A.1 Parameters of Asynchronous Machines

$$k_2 = \left(\frac{24}{16} \frac{y}{\tau_p} - \frac{4}{16} \right) \qquad \frac{1}{3} \leq \frac{y}{\tau_p} \leq \frac{2}{3}, 60° \text{ zones}, 3 - \text{phases}, \qquad (A.21)$$

$$k_2 = \left(\frac{12}{16} \frac{y}{\tau_p} + \frac{4}{16} \right) \qquad 0 \leq \frac{y}{\tau_p} \leq \frac{1}{3}, 60° \text{ zones}, 3 - \text{phases}, \qquad (A.22)$$

$$w = n_1 q 2 p$$

with

w: number of stator windings per phase,
n_1: number of conductor per half slot
a: number of parallel circuits in the stator windings w
l_{act}: length of active part
l_{Fe}: iron length,
r_i: inner stator radius,
h_{11}: Twice the height of one winding bar,
d_1: height between upper and lower bar,
h_{12}: stator leakage height,
A_{Cu}: Copper cross-section of one strand
δ: air gap width enlarged by the carter factor of rotor and stator,
p: number of pole pairs,
Z_2: number of rotor slots,
$^p k_w$: winding factor,
q: number of slots per pole and phase
y: Winding width
τ_p: Pole width
r_r: radius of cage ring
l_r: length of the cage ring between two bars

A.2 Parameters of Synchronous Machines

Main synchronous machine parameters are given by the stator and the cage quantities of the asynchronous machine. Additionally the field winding resistance R_f and the main inductivity $M_{f,1}$ have to be defined.

$$R_f = \frac{w_f}{a_{f2}^2} \cdot \frac{2(l_{Fe,f} + l_{S,f})}{\gamma_f A_{cu,f}}, \tag{A.23}$$

$$M_{1,f} = \frac{2}{\pi} \cdot \frac{\left(\dfrac{w_f}{p}\right) k_{wf} k_{w1} \cdot q \cdot z_n}{a \cdot a_{f2}} \cdot \frac{\mu_0 \cdot 2r \cdot l_{Fe}}{\delta} \tag{A.24}$$

$$z_n = 2 \cdot n_1, \tag{A.25}$$

$$w_f = p \sum_{i=1}^{q_f} n_{fi}, \tag{A.26}$$

$$k_{wf} = \frac{1}{q_f} \sum_{i=1}^{q_f} \frac{n_{fi}}{n_m} \sin(p \alpha_i), \tag{A.27}$$

$$n_m = \frac{w_f}{p q_f}, \tag{A.28}$$

with

w_f: total number of field windings,
a_{f2}: number of parallel circuits in the exciter winding,
$l_{Fe,f}$: iron length-rotor,
$l_{S,f}$: end winding length-field winding,
$A_{Cu,f}$: Copper cross-section of one conductor in the field winding
q_f: number of exciter slots per pole,
n_{fi}: number of conductors in slot i,
n_m: mean number of conductors per slot
$\alpha_{f,i}$: distance between the middle of the slot i and the pole axis in rad,
k_{wf}: winding factor of the field winding,
n_1: number of conductors per stator half slot,
z_n: number of conductors per stator slot.

Symbols

Chapter 2

Symbol	Unit	Description
I_D	(A)	d. c. load current
I_p	(A)	primary effective transformer current
I_s	(A)	secondary effective transformer current
S_{Tr}	(VA)	rated power of converter transformer
U_A	(V)	load voltage
U_d	(V)	d.c. link voltage
$U_{dc,bef}$	(V)	d.c. link voltage before the choke
$U_{dc,aft}$	(V)	d.c. link voltage after the choke
U_{Di}	(V)	d.c. load voltage of the inner ohmic resistance
U_p	(V)	primary effective transformer voltage
U_{RS}	(V)	phase to phase grid voltage
U_s	(V)	secondary effective transformer voltage
α	(rad)	thyristor delay angle

Chapter 3

Symbol	Unit	Description
$a_{cs+,w}$	(m^2)	cross-section within the positive current flow direction
$a_{cs-,w}$	(m^2)	cross-section within the negative current flow direction
A_z	(Wb/m)	magnetic vector-potential in z-direction
D_i		element with diode circuit characteristic
E	(V)	Induced voltage in the armature winding due to the excitation field
e_z		unity vector in z-direction
$F_{\gamma,\iota,k}$	(rad S)	conductance coefficient of the FD-calculation
$F_{w,I}$	(1/m^2)	winding density factor
h	(s)	distance in time between two time steps

Symbol	Unit	Description
I_f	(p.u.)	field winding current
I_r	(A)	diode reverse current
$i_{w,i}$	(A)	current within the winding i
$J_{i,k}$	(A/m^2)	current density in the mesh element between the k-th and k+1-th and the i-th and i+1-th line
$\underline{I}_1$	(A)	complex stator current in one pole winding
$\underline{I}_{R,2}$	(A)	complex rotor current in the rotor end ring
$\underline{I}_{St,2}$	(A)	complex rotor current in one rotor bar
k	(V/T)	Boltzmann constant
k_w		winding factor
L_{n1}	(H)	stator slot leakage,
L_{S1}	(H)	stator end winding leakage
$L_{1\sigma}$	(H)	stator leakage
$L_{2\sigma}$	(H)	rotor leakage
L_1	(H)	stator main inductivity
L_2	(H)	rotor main inductivity
$M_{1,f}$	(H)	maximum mutual inductivity between the armature- and the field winding
L_{N2}	(H)	rotor slot leakage,
L_{R2}	(H)	rotor end winding leakage,
$L_{\sigma w}$	(H)	leakage inductance of the winding
m		number of phases
$M_{1,2}$	(H)	mutual inductivity stator - rotor
$M_{2,1}$	(H)	mutual inductivity rotor - stator
M_{asyn}	(Nm)	torque of an asynchronous machine
m_e		diode emission constant
M_S	(Nm)	stalling torque of an asynchronous machine
N, S, M, O, W	(rad A/Tm)	reluctance coefficients of the FD-calculation
$N_{turn\,w}$		number of turns of the winding
p		number of pole pairs
$P_{cu,1}$	(W)	stator copper losses
$P_{cu,2}$	(W)	rotor copper losses
p_i	(rad)	distance between the i-th and i+1-th radial mesh line
q_k	(m)	distance between the k-th and k+1-th tangential mesh line

Symbols

Symbol	Unit	Description
R_1	(Ω)	stator resistance
R_f	(p.u.)	resistance of the field winding
r_k	(m)	radius of the k-th tangential mesh line
R_{R2}	(Ω)	resistance in the rotor end ring
R_{S2}	(Ω)	resistance in the rotor bar
R_Q	(p.u.)	resistance of the damper winding in the q-axis
R_w	(Ω)	resistance of the winding
s		slip
$\underline{U}_1$	(V)	complex stator phase voltage of one pole winding
u_d	(V)	diode voltage
U_f	(p.u.)	voltage of the field winding
U_T	(V)	thermo-diode voltage
$u_{w,i}$	(V)	voltage within the winding i
w_1		number of stator windings of one phase
w_2		number of rotor windings, in case of a rotor cage it is one
Z	(Ω)	impedance
Z_2		number of rotor bars
χ	(S/m)	conductivity
σ_{d1}	(p.u.)	air-gap leakage flux related to the main stator inductance
σ_{d2}	(p.u.)	air-gap leakage flux related to the main rotor inductance
φ	(V/m)	electrical potential
φ_i	(rad)	angular coordinate of the i-th radial mesh line
μ	(Tm/A)	permeability
$\Psi_{w,i}$	(Wb)	flux linkage within the winding i
Θ		factor, for an increased numerical stability of the time integration
ω	(s^{-1})	angular frequency
ω_{el}	(s^{-1})	electrical angular frequency
ω_{mec}	(s^{-1})	mechanical angular frequency

Chapter 4

Symbol	Unit	Description
a	(1/m)	reciprocal of skin depth
A_s	(m^2)	loss surface
$\underline{A_z}$	(Wb/m)	complex vector potential
$^1\hat{B}_{0,\delta}$	(T)	radial flux density air-gap amplitude of the fundamental static distribution without damping effects
b_1	(m)	conductor width
$^1\hat{B}_{d,\delta}$	(T)	radial flux density air-gap amplitude of the fundamental distribution with damping effects
B_m	(T)	mean air-gap flux density
b_n	(m)	slot width
C_{Fe}	(Ws/T)	material specific loss constant for iron losses
E_a	(V/m)	external field strength
f_{el}	(Hz)	electrical frequency of the air-gap ripple
h	(m)	strand height
$\hat{H}_\theta$	(A/m)	tangential component of the magnetic field strength amplitude
$\underline{J_z}$	(A/m^2)	complex current density in axial direction
k_{Fe}		iron loss factor for higher harmonics
k_{inter}		influence of the harmonic interaction on iron losses
k_L		harmonic loss factor
k_{rtl}		factor for the resistance increase in strands with the same current
k_{rut}		factor for the resistance increase in strands with parallel connected parts
l_{ges}	(m)	total copper length including cooling ducts and end winding regions
L_{n10}	(H)	stator slot leakage for d.c. current distribution
n		number of conductors in a slot
$P_{1,0}$	(W)	no load stator losses
$P_{1,r}$	(W)	rated power at the motor terminals
$P_{cu,1,0,dc}$	(W)	calculated no load copper d.c. losses in the stator, which are calculated from the no load current
P_{cu2}	(W)	rotor losses

Symbols

Symbol	Unit	Description
$P_{\text{cu1,dc}}$	(W)	copper stator losses without eddy currents
P_{Fe}	(W)	magnetization losses or iron losses including higher harmonic no load currents
P_{Reib}	(W)	friction losses
P_{ri}	(W)	losses due to air-gap flux ripple
P_{vw}	(W)	copper losses, taking eddy currents into account
P_{δ}	(W)	air-gap power transmission
$R_{1,\text{dc,r}}$	(Ω)	stator d.c. resistance at rated operation temperature
$R_{1,\text{dc,m}}$	(Ω)	measured stator d.c. resistance
R_{n0}	(Ω)	strand resistance of one strand
s		slip
S_{dr}	$\left(\sqrt{\text{Hz S}/\text{m}}\;\text{T m}\right)$	saturation factor
z		number of parallel strand parts within one strand
α_1		relationship between the fundamental flux density ripple and the mean air-gap flux density
χ	(S/m)	conductivity
φ_{ph}	(rad)	phase shift between upper and lower layer in a slot
γ	(1/m)	skin depth parameter
ϑ_{m}	(°C)	stator temperature during the measurement
ϑ_{r}	(°C)	stator temperature at rated operation
λ	(m)	pole pitch of the air-gap ripple
ν		harmonic number
$\mu_{\text{r,ob}}$		relative equivalent permeability according Oberretl's theorie
Ψ		additional loss contribution of adjacent strands
Ψ'		slot leakage inductance decrease due to currents in adjacent strands
Ψ_1	(Wb)	flux linkage of the fundamental
Ψ_ν	(Wb)	flux linkage of the ν harmonic
ς		resistance increase of a single strand
ς'		slot leakage inductance decrease of a single strand

Chapter 5

Symbol	Unit	Description
D_1	(s Nm)	damping coefficients
h	(m)	cylinder height
I_P	(m^4)	polar geometrical moment of inertia
J_1	(kg m^2)	shaft inertia moments
$J_{\bar{x}}$	(kg m^2)	shaft inertia moment for a rotation around the center of gravity
J_x	(kg m^2)	shaft inertia moment for a rotation around the x-axis
J_y	(kg m^2)	shaft inertia moment for a rotation around the y-axis
J_z	(kg m^2)	shaft inertia moment for a rotation around the z-axis
$k_{1,k}$	(Nm)	torsional stiffness coefficients
k_t	(Nm)	torsional stiffness
M	(Nm)	electromagnetic torque
m	(kg)	rotating mass
q		modal coordinates
r	(m)	outer cylinder radius
U_{m1}, U_{m2}		converter harmonics
W_m	(Ws)	magnetical energy
(X)		matrix of Eigenvectors of the mechanical structure
(X_T)		transposed of the matrix
(X^1)		conjungate of the matrix
β_2	(rad)	virtual rotor rotation
Λ		logarithmic decrement
δ		decay constant
Λ_1, Λ_2		permeance functions
ϑ_r	(°C)	stator temperature at rated operation
ϑ_m	(°C)	measured stator temperature
ρ	(kg/m^3)	material density
τ		shearing stress
τ_{max}		maximal shearing stress
ω_1	(s^{-1})	angular frequency of the stator field within the stator coordinate system
ω_2	(s^{-1})	angular frequency of the rotor field within the rotor coordinate system
$\omega_{el,netz}$	(s^{-1})	electrical angular grid frequency
ω_{mec}	(s^{-1})	mechanical angular frequency of the shaft

Symbols 175

Chapter 6

Symbol	Unit	Description
$\hat{B}_{-,\nu,\mu,\cos}$	(T)	cosine amplitude of the air-gap flux density in the negative traveling direction for the ν-th space harmonic and μ-th time harmonic
$\hat{B}_{+,\nu,\mu,\cos}$	(T)	cosine amplitude of the air-gap flux density in the positive traveling direction for the ν-th space harmonic and μ-th time harmonic
c	(m/s)	sonic speed
C'	(N/m)	stiffness per surface for the air-gap
d	(m)	thickness of the lamination
E	(N/m^2)	elasticity modulus
F'	(N/m^2)	magnetic force density
$\hat{f}_{-,\nu,\mu,\cos}$	(N/m^2)	cosine amplitude of the air-gap force density in the negative traveling direction for the ν-th space harmonic and μ-th time harmonic
$\hat{f}_{+,\nu,\mu,\cos}$	(N/m^2)	cosine amplitude of the air-gap force density in the positive traveling direction for the ν-th space harmonic and μ-th time harmonic
h_y	(m)	stator yoke height
$\vec{n}$		vector perpendicular to the surface
r		mechanical mode
R_m	(m)	radius towards the middle of the stator yoke
R_δ	(m)	bohre radius
T_m	(N/m^2)	magnetic stress vector
$\vec{x}$	(m)	position vector in space
Δd	(m)	elongation due to magnetic forces
ρ	(kg/m^3)	density

Chapter 7

Symbol	Unit	Description
a		number of parallel circuits
C_a	(F)	capacity of one stator phase
C_b	(F)	capacity over the bearings
C_{rf}	(F)	capacity between rotor and stator frame
C_{rw2}	(F)	capacity between rotor winding and rotor
C_{wf}	(F)	capacity between stator winding and stator frame
C_{wr}	(F)	capacity between stator winding and rotor
d	(m)	strand depth
d_{se}	(m)	stator outer diameter
d_{si}	(m)	bore diameter
h_s	(m)	slot height
k_i		reduction in the leakage inductance due to eddy currents
$l_{fe,core}$	(m)	iron length of the stator joke
$l_{fe,tooth}$	(m)	iron length of a stator tooth
L_g	(H)	mutual inductivity between stator ground current and shaft
$L_{\sigma,bar}$	(H)	leakage inductance in the area of the bar cross-section
$L_{\sigma,s}$	(H)	leakage inductance in the area of the slot opening
n		number of chain conductor elements per phase and per parallel circuit
N_{Fe}		number of laminations
R_1	(Ω)	resistance of winding bar
R_{ac}	(Ω)	a.c.-resistance of the bar
R_{acsT}	(Ω)	a.c.-resistance of one strand
R_{dc}	(Ω)	d.c.-resistance
r_m	(m)	mean radius of the stator core
u_b	(V)	voltage over the bearings between ground and shaft
u_{cm}	(V)	common mode voltage
δ_s	(m)	skin depth in space
$\delta_{T,1}$	(s)	skin depth in time for a single strand
$\delta_{T,2}$	(s)	skin depth in time for adjacent strands
$\delta_{T,bar}$	(s)	skin depth in time for a bar with several strands

Symbols

Chapter 8

Symbol	Unit	Description
C'	(F/m)	capacity per conductor length
C_{rf}	(F)	capacity between the rotor and the ground potential of the core
C_{tbe}	(F)	capacity between the bottom turn and the ground potential of the core
C_{te}	(F)	capacity between a turn and the ground potential of the core
C_{tt}	(F)	capacity between two turns
C_{wf}	(F)	capacity between a winding layer and the ground potential of the core
C_{wr}	(F)	capacity between the lower winding layer and the rotor
C_{ww}	(F)	capacity between two winding layers
G'	(S/m)	conductance per conductor length
l	(m)	length of the line
L'	(H/m)	inductance per conductor length
Q		Eigen-vectors of the matrix product of impedance and admittance
R'	(Ω/m)	resistance per conductor length
t_a	(s)	surge rise time
t_h	(s)	pulse half time
$[Y]$	(S)	admittance matrix of the line
$[Y_0]$	(S)	characteristic admittance matrix
Z	(Ω)	motor impedance
Z_0	(Ω)	cable impedance
$[Z]$	(Ω)	impedance matrix of the line
γ		Eigen-value of the matrix product of impedance and admittance

Chapter 9

Symbol	Unit	Description
n	(rpm)	speed
P_{loss}	(W)	process and motor rotor losses
P_{pump}	(W)	required pump power incl. pump losses
P_δ	(W)	air-gap power
$\dot{V}$	(m^3/s)	volume flow
$\dot{V}_r$	(m^3/s)	rated volume flow
η_{conv}		converter efficiency
η_{gear}		gear efficiency
η_{motor}		motor efficiency

Attachment

Symbol	Unit	Description
a		number of parallel stator windings per phase
$A_{Cu,a}$	(m^2)	copper cross-section of one bar in the armature
$A_{Cu,D}$	(m^2)	sum of copper cross-section of all damper bars per pole
$A_{Cu,f}$	(m^2)	copper cross-section of one conductor in the field winding
a_{f2}		number of parallel circuits in the exciter winding
d_1	(m)	height between upper and lower bar
h_{11}	(m)	twice the height of one winding bar
h_{12}	(m)	stator leakage height
k_w		winding factor
$l_{Fe,a}$	(m)	iron length-armature winding
$l_{Fe,f}$	(m)	iron length-rotor
l_r		cage-ring-length between two bars
$l_{S,a}$	(m)	end winding length-armature
$l_{S,f}$	(m)	end winding length-field winding,
q		number of slots per pole and phase
q_f		number of exciter slots per pole
r_i	(m)	inner stator radius
r_r	(m)	radius of cage ring
w_a		number of armature windings per phase
w_f		total number of field windings
$n_{f,i}$		number of conductors in slot i
z_n		number of conductors per stator slot

Symbols

Symbol	Unit	Description
$\alpha_{f,i}$	(rad)	distance between slot i and the pole axis in rad
δ_i	(m)	ideal air-gap width enlarged by the carter factor of rotor and stator
ϑ	(rad)	polar wheel angle
y		winding width
τ_p		pole width

Abbreviations

VSI	voltage source inverter
LCI	load commutated current source inverter
IGBT	Insulated-gate bipolar transistor
FEM	finite element method
FD	finite difference
DE	driven end
NDE	non-driven end
EMI	Electromagnetic interference
DCS	distributed control system
CSI	current source inverter
HF	high frequency

References

Chapter 1

[1.1] Reprint, N.N.: Repowering Project for EnBW Kraftwerk AG using GT26 advanced gas turbine in Karlsruhe Rheinhafen, Modern Power Systems (April 2002)

[1.2] Drubel, O.: Static Frequency converters with Reduced Parasitic Effects. In: IEEE PESC 2004, Aachen (2004)

[1.3] International Electrotechnical Comission, IEC 60034, Rotating Electrical Machines

[1.4] American Petroleum Institute, API STD 670, Machinery Protection Systems

[1.5] Shell, Shell DEP 33.66.05.31

[1.6] ZLM, Zusätzliche Lieferbedingungen für Motoren in Kraftwerken

[1.7] Saudi Aramco Materials System Specification, 17-SAMSS-503, Severe-Duty, Totally Enclosed, Squirrel Cage Induction Motors to 500HP (370kW)

[1.8] Oberretl, K.: Losses, torques and magnetic noise in induction motors with static converter supply, taking multiple armature reaction and slot openings into account. IET Electr. Power Appl. 1(4), 517–531 (2007)

[1.9] Joksimovic, G., Binder, A.: Additional no-load losses in inverter fed high-speed cage induction motors. Electrical Engineering (86), 105–116 (2004)

[1.10] Drubel, O.: Converter Dependent Design of Induction Machines in the Power Range below 10MW. In: Proceedings of IEEE IEMDC, SS 5.2, Antalya (2007)

[1.11] Miliani, E.H., Depernet, D., Kauffmann, J.M., Lacaze, A.: Experimental control of matrix converter for active generator. In: IEEE PESC 2004, Aachen, pp. 2899–2904 (2004)

[1.12] Drubel, O.: Erweitertes Verständnis elektromagnetischer Vorgänge und Temperaturverteilungen im Rotor großer Turbogeneratoren für gestörten Betrieb durch Ableitung und Einsatz Finite-Differenzen- Zeitschrittrechnung, Dissertation Universität Dortmund (2000)

[1.13] Neidhöfer, G.: Michael von Dolivo-Dobrowolsky und der Drehstrom – Anfänge der modernen Antriebstechnik und Stromversorgung. VDE Verlag GmBH, Berlin (2004)

[1.14] Allgemeine Elektrizitäts Gesellschaft in Berlin (Doliwo-Dobrowolski), Deutsches Patent 51083, Anker für Wechselstrommotoren, angemeldet am 8.3.1889, erteilt am 19.4.1890 (1889)

[1.15] Drubel, O.: Future challenges within numerical field calculation for industrial machines. In: CEM 2006, Aachen, pp. 21–24 (2006)

Chapter 2

[2.1] Kulig, T.S.: Über die Auswirkungen von Störfällen in elektrischen Energieübertragungsnetzen auf Kraftwerksturbosätzen. Habilitation, Fernuniversität Hagen (1987)

[2.2] Hammons, T.J., Goh, M.W.: Turbine, Generator, System Modeling and Impact of Variable Frequency Ripple Currents on Torsional Stressing of Generators in Poland and Sweden: Lithuania/Poland and Sweden/Poland HVDC Links. IEEE Transactions on Energy Conversion 15(4), 384–394 (2000)

[2.3] Bird, B.M., King, K.G., Pedder, D.A.G.: An Introduction to Power Electronics, 2nd edn. John Wiley & Sons, Chichester (1993)

[2.4] Zach, J.: Leistungselektronik: Bauelemente, Leistungskreise, Steuerungskreise, Beeinflussungen, 3., verb. u. erg. Aufl. – Wien. Springer, New York (1990)

[2.5] Meyer, M.: Leistungselektronik: Einführung, Grundlagen, Überblick. Springer, Heidelberg (1990)

[2.6] Kümmel, F.: Elektrische Antriebstechnik. Springer, Heidelberg (1971)

[2.7] Schröder, D.: Elektrische Antriebe 4 Leistungselektronische Schaltungen. Springer, Heidelberg (1998)

[2.8] Kassakian, J.G., Schlecht, M.F., Verghese, G.C.: Principles of Power Electronics. Addison-Wesley Publishing Company (1991)

[2.9] McGrath, B.P., Holmes, D.G., Lipo, T.: Optimized Space Vector Switching Sequences for Multilevel Inverters. IEEE Trans. on Power Electronics 18(6), 1293–1301 (2003)

[2.10] Peng, F.Z., Lai, J.S., McKeever, J.W., Van Coevering, J.: A multi-level voltage-source inverter with separate DC sources for static var generation. IEEE Trans. on Industry Application 32, 1130–1138 (1996)

[2.11] Manjrekar, M.D., Steimer, P., Lipo, T.: Hybrid multi-level power conversion system: A competitive solution for high power applications. Trans. on Industry Applications 36, 834–841 (2000)

[2.12] Lai, J., Peng, F.: Multi-level converters – A new breed of power converters. Trans. on Industry Applications 32, 509–517 (1996)

[2.13] Jenni, F., Wüest, D.: Steuerverfahren für selbstgeführte Stromrichter. vdf Hochschulverlag AG an der ETH Zürich und B.G. Teubner Stuttgart (1995)

Chapter 3

[3.1] Oberretl, K.: Berechnung des Streuflusses im Luftspalt von elektrischen Maschinen mit Käfig- oder Dämpferwicklung, Teil 1: Theorie und Berechnungsmethoden. Electrical Engineering (AfE) 69, 11–22 (1986)

[3.2] Oberretl, K.: Die genauere Berechnung des Magnetisierungsstromes von dreiphasigen Asynchronmaschinen. Bulletin Oerlikon (335), 66–84 (1959)

[3.3] Schuisky, W.: Induktionsmaschinen. Springer, Wien (1957)

[3.4] Canay, M.: Ersatzschemata der Synchronmaschine sowie Vorrausberechnung der Kenngrössen mit Beispielen, Dissertation, Ecole Polytec. de l'Universite de Lausanne (1968)

[3.5] Dommel, H.: Digitale Rechenverfahren für elektrische Netze, Dissertation, Technische Hochschule München (1962)

[3.6] Sapin, A., Simond, J.J.: Simsen: A modular software package for the analysis of power networks and electrical machines. In: CICEM 1995, Hangzhou, China (1995)

References   183

[3.7] Drubel, O., Lacaze, A., Karachev, A.: Enhanced electrical systems by voltage and frequency controlled brushless excitation. In: Proceedings of ICEM 2004, Krakow (2004)

[3.8] Simony, K.: Theoretische Elektrotechnik. VEB Deutscher Verlag der Wissenschaften, Berlin (1977)

[3.9] Simond, J.J.: Simsen Software (2004), http://Simsen.epfl.ch

[3.10] Bortsov, Y., Polyakhov, N., Loginov, A., Burmistrov, A.: Robust Controller for Excitation Systems of Synchronous Generators. In: International Conference on Electrical Machines (ICEM 2000), Espoo, Finland, August 28-30, vol. 2(3), pp. 1056–1060 (2000)

[3.11] Kitauchi, Y., Taniguchi, H., Shirasaki, T., Ichikawa, Y., Amano, M., Banjo, M.: Experimental Verification of Multi-input PSS with Reactive Power Input for Damping Low Frequency Power Swing. IEEE Trans. on Energy Conversion 14(4), 1124–1130 (1999)

[3.12] Abido, M.A., Abdel-Magid, Y.L.: Optimal Design of Power System Stabilizers Using Evolutionary Programming. IEEE Trans. on Energy Conversion 17(4), 429–436 (2002)

[3.13] Reichert, K.: Über ein numerisches Verfahren zur Berechnung von Magnetfeldern und Wirbelströmen in elektrischen Maschinen. Habilitation Universität Stuttgart (1968)

[3.14] Gottkehaskamp, R.: Nichtlineare Berechnung von Asynchronmaschinen mit massiveisernem Rotor und zusätzlichem Käfig im transienten Zustand mittels Finiter Differenzen und Zeitschrittrechnung, PhD thesis, Univ. Dortmund, VDI-Verlag, VDI-Fortschrittbericht, Nr. 131 (1993)

[3.15] Klocke, M.: Determination of Dynamical Phenomena in Soft-Started Induction-Motor-Drives using the Finite-Difference-Time-Stepping Method. In: Proceedings of IX International Symposium on Electromagnetic Fields in Electrical Engineering (ISEF 1999), Pavia, Italy, September 23-25 (1999)

[3.16] Drubel, O.: Elektromagnetische Vorgänge und Temperaturverteilungen im Rotor grosser Turbogeneratoren im gestörten Betrieb, Fortschr.-Ber. VDI Reihe 21 Nr. 304. VDI Verlag, Düsseldorf (2001)

[3.17] Tsukerman, I.A., Konrad, A., Meunier, G., Sabonnadiere, J.C.: Coupled field-circuit problems: Trends and accomplishments. IEEE Transactions on Magnetics 29(2), 1701–1704 (1993)

[3.18] Biddlecombe, C.S., Simkin, J., Jay, A.P., Sykulski, J.K., Lepaul, S.: Transient electromagnetic analysis coupled to electric circuits and motion. IEEE Transactions on Magnetics 34(5), 3182–3185 (1998)

[3.19] Demenko, A.: Movement simulation in finite element analysis of electric machine dynamics. IEEE Transactions on Magnetics 32(3), 1553–1556 (1996)

[3.20] Drubel, O.: Current distribution within multi strand windings for electrical machines with frequency converter supply. In: Proceedings of ICEM 2002, Brugge (2002)

[3.21] De Gersem, H., Mertens, R., Lahaye, D., Vandewalle, S., Hameyer, K.: Solution strategies for transient, field-circuit coupled systems. IEEE Transactions on Magnetics 36, 1531–1534 (2000)

[3.22] De Gersem, H., Mertens, R., Pahner, U., Belmans, R., Hameyer, K.: A topological method used for field-circuit coupling. IEEE Transactions on Magnetics 34, 3190–3193 (1998)

References

[3.23] Kost, A.: NumerischeMethoden in der Berechnung elektromagnetrischer Felder. Springer, Berlin (1994)

[3.24] Darabi, A., Tindall, C., Ferguson, S.: Finite Element Time –Step Coupled Generator, Load, AVR, and Brushless Exciter Modelling. IEEE Trans. Energy Conversion 19(2), 258–264 (2004)

[3.25] Darabi, A., Tindall, C.: Damper Cages in Genset Alternators: FE Simulation and Measurement. IEEE Trans. Energy Conversion 19(1), 73–80 (2004)

[3.26] Drubel, O., Gantenbein, R., Izquierdo, A., Klocke, M.: Current flow and losses in brushless exciters with polygon-connected windings and dc rectifiers. Electrical Engineering (AfE) 90(1) (November 2007)

[3.27] Mathis, W.: Theorie nichtlinearer Netzwerke. Springer, Heidelberg (1987)

[3.28] Klocke, M.: Zur Berechnung dynamischer Vorgänge bei von einem Drehstromsteller gespeisten Antrieben mit Asynchronmaschinen und mehreren gekoppelten Massen mittels Finite-Differenzen-Zeitschritt-Rechnung, Dissertation Universität Dortmund (1999)

Chapter 4

[4.1] International Electrotechnical Comission, IEC 60034-2, Rotating Electrical Machines

[4.2] Drubel, O.: Current distribution within multi strand windings for electrical machines with frequency converter supply. In: Proceedings of ICEM, Brugge, no. 627 (2002)

[4.3] Müller, G., Vogt, K., Ponick, B.: Berechnung elektrischer Maschinen. Wiley-VCH-Verlag, Weinheim (2008)

[4.4] Haldemann, J.: Transpositions in stator bars of large turbogenerators. IEEE Transactions on Energy Conversion 19(3), 553–560 (2004)

[4.5] Oberretl, K.: Eisenverluste, Flußpulsation und magnetische Nutkeile in Käfigläufermotoren. Electrical Engineering 82, 301–311 (2000)

[4.6] Lancarotte, M.S., de Penteado, A.: Estimation of Core Losses under Sinusoidal or Non-Sinusoidal Induction by Analysis of Magnetization Rate. IEEE Transactions on Energy Conversion 16(2), 174–179 (2001)

[4.7] Bozorth, R.M.: Ferromagnetism. IEEE Press, New York (1993)

[4.8] Oberretl, K.: Eddy current losses in solid pole shoes of synchronous machines at no-load and on load. IEEE Trans. PAS 91, 152–160 (1972)

[4.9] Gibbs, W.J.: Tooth-ripple losses in unwound pole-shoes. Journal IEE (London) 94(pt. II), 2–12 (1947)

[4.10] Lawrenson, P.J., Reece, P., Ralph, M.C.: Tooth-ripple losses in solid poles. Proceedings IEE (London) 113, 657–662 (1966)

[4.11] Gottkehaskamp, R.: Nichtlineare Berechnung von Asynchronmaschinen mit massiveisernem Rotor und zusätzlichem Käfig im transienten Zustand mittels Finiter Differenzen und Zeitschrittrechnung. VDI Fortschrittsberichte, Düsseldorf (1993)

[4.12] Oberretl, K.: Die genaue Berechnung des Magnetisierungsstromes von dreiphasigen Asynchronmaschinen. Bulletin Oerlikon Nr.335, S.66 (1959)

[4.13] Reichert, K.: FEMAG, Finite Elemente Calculation program, ETH-Zürich (1999)

[4.14] Harris, M.R., Fam, W.Z.: Analysis and measurement of radial power flow in machine air gaps. Proceedings IEE 113(10), pp.1607–1615 (1966)

References

[4.15] Drubel, O., Stoll, R.L.: Comparison between Analytical and Numerical Methods of Calculating Tooth Ripple Losses in Salient Pole Synchronous Machines. IEEE Trans. Energy Conversion 16(1), 61–67 (2001)

[4.16] Cerovsky, Z., Seinsch, H.O.: Zusätzliche Stromwärmeverluste im Läufer von Induktionsmaschinen und ihre räumliche Verteilung bei Spei-sung mit U-Umrichtern im Blockbetrieb. Archiv für Elektrotechnik (AfE) 77, 197–202 (1994)

[4.17] Cerovsky, Z., Seinsch, H.O.: Verlauf und Symmetrieeigenschaften der Läuferströme von Induktionsmotoren bei Speisung mit U-Umrichtern im Blockbetrieb. Archiv für Elektrotechnik (AfE) 77, 107–115 (1994)

Chapter 5

[5.1] Oberretl, K.: Parasitäre synchrone Dreh- und Pendelmomente in Asynchronmotoren, Einfluß von Ausgleichsvorgängen und Eisensättigung. Teil I, Archiv für Elektrotechnik (AfE) 77, 179–190 (1994)

[5.2] Oberretl, K.: Parasitäre synchrone Dreh- und Pendelmomente in Asynchronmotoren, Einfluß von Ausgleichsvorgängen und Eisensättigung, Teil II. Archiv für Elektrotechnik (AfE) 77, 277–288 (1994)

[5.3] Oberretl, K.: Allgemeine Oberfeldtheorie für ein- und dreiphasige Asynchron- und Linearmotoren mit Käfig unter Berücksichtigung der Nutöffnungen Teil I: Theorie und Berechnungsverfahren. Archiv für Elektrotechnik (AfE) 76, 111–120 (1993)

[5.4] Oberretl, K.: Allgemeine Oberfeldtheorie für ein- und dreiphasige Asynchron- und Linearmotoren mit Käfig unter Berücksichtigung der Nutöffnungen Teil II: Resultate, Vergleich mit Messungen. Archiv für Elektrotechnik (AfE) 76, 203–212 (1993)

[5.5] Wagner, W.: Berechnung von Drehstromasynchronmaschinen mit Käfigläufern unter Berücksichtigung von mehrfacher Ankerrückwirkung, Nutöffnungen und Rotorquerströmen, PhD. thesis, Univ. Dortmund (1986)

[5.6] Oberretl, K.: Losses, torques and magnetic noise in induction motors with static converter supply, taking multiple armature reaction and slot opening into account. IET Electr. Power Appl. 1(4), 517–531 (2007)

[5.7] Kulig, T.S.: Über die Auswirkungen von Störfällen in elektrischen Energieübertragungsnetzen auf Kraftwerksturbosätzen, Habilitation, Fernuniversität Hagen (1987)

[5.8] Beitz, W., Grote, K.-H., et al.: Dubbel Taschenbuch für den Maschinenbau, 19. Afl. Springer Verlag (1997)

[5.9] Dresig, H.: Schwingungen mechanischer Antriebsysteme – Modellbildung, Berechnung, Analyse, Synthese, 1.Afl. Springer, Heidelberg (2005)

[5.10] Sähn, S.: Torsionsfederzahlen abgesetzter Wellen mit Kreisquerschnitt und Folgerungen für die Gestaltung von Schrumpfverbindungen. Konstruktion 19, Heft 1 S.12–S.19 (1967)

[5.11] Jarausch, R., Mader, H.: Berechnung erzwungener gedämpfter Drehschwingungen von Getrieben mit Hilfe elektronischer Rechenmaschinen. Industrie-Anzeiger 84(63), S.253–S.262 (1962)

[5.12] Seinsch, H.O.: Oberfelderscheinungen in Drehfeldmaschinen: Grundlagen zur analytischen und numerischen Berechnung. Teubner Verlag, Stuttgart (1992)

[5.13] Oberretl, K., Scho, C., Wagner: Programm Amoto, TU Dortmund (1995)

[5.14] Brandes, J.: Beanspruchung des Wellenstranges bei umrichtergespeisten Asynchronmaschinen. Archiv für Elektrotechnik (AfE), 115–130 (1990)

Chapter 6

[6.1] Oberretl, K.: Losses, torques and magnetic noise in induction motors with static converter supply, taking multiple armature reaction and slot opening into account. IET Electric Power Applications 1(4), 517–531 (2007)

[6.2] Haaf, D.: Einfluss des Umrichterbetriebs auf das akustische Verhalten im Gesamtsystem Asynchronmotor-Getriebe, PhD. Thesis, RWTH Aachen (2004)

[6.3] Jordan, H., Greiner, M.: Mechanische Schwingungen. Giradet Verlag, Essen (1952)

[6.4] Seinsch, H.O.: Oberfelderscheinungen in Drehfeldmaschinen: Grundlagen zur analytischen und numerischen Berechnung, Stuttgart, Teubner Verlag (1992)

[6.5] Aschendorf, B.: Zur Berechnung des magnetischen Geräusches von Käfigläufermotorn, PhD thesis, Universität Dortmund (1990)

[6.6] Lach, R.: Magnetische Geräuschemission umrichtergespeister Käfigläufer Asynchronmaschinen, PhD thesis, TU Dortmund (2006)

[6.7] Drubel, O., Gantenbein, R., Izquierdo, A., Klocke, M.: Current flow and losses in brushless exciters with polygon-connected windings and dc rectifiers. Electrical Engineering (AfE) 90(1) (November 2007)

[6.8] Klocke, M.: Zur Berechnung dynamischer Vorgänge bei von einem Drehstromsteller gespeisten Antrieben mit Asynchronmaschinen und mehreren gekoppelten Massen mittels Finite-Differenzen-Zeitschrittrechnung, Dissertation Universität Dortmund (1999)

[6.9] Frohne, H.: Über die primären Bestimmungsgrößen der Lautstärke bei Asynchronmaschinen, PhD thesis, TH Hannover (1959)

[6.10] Trochidis, A.: Lärmarm konstruieren (IV) – Körperschalldämpfung mittels Gas- und Flüssigkeitsschichten, Forschungsbericht Nr. 203, im Auftrag des Bundesministers für Arbeit und Sozialordnung (1979)

[6.11] Cremer, L., Heckel, M.: Körperschall Physikalische Grundlagen und technische Anwendungen. Springer Verlag, 2. Auflage (1996)

[6.12] Habetler, T., Deepakraj, M.: Acoustic Noise Reduction in Sinusoidal PWM Drives Using a Randomly Modulated Carrier. IEEE Transactions on Power Electronics 6(3), 356–363 (1991)

Chapter 7

[7.1] Mütze, A.: Bearing currents in Inverter-Fed AC-Motors, Berichte aus der Elektrotechnik. Shaker Verlag, Aachen (2004)

[7.2] Drubel, O.: Static Frequency converters with Reduced Parasitic Effects. In: IEEE PESC 2004, Aachen, pp. 4365–4370 (2004)

[7.3] Amann, C., Reichert, K., Joho, R., Possedel, Z.: Shaft Voltages in Generators with Static Excitation Systems – Problems and Solution. IEEE Trans. Energy Conversion 3(2) (June 1988)

[7.4] Zitzelsberger, J., Stupin, P., Hofmann, W.: Bearing Currents in Doubly-Fed Induction Generators. In: EPE 2005, Dresden, September 2005, p. 9 (2006)

[7.5] Torlay, J.E., Corenwinder, C., Audoli, A., Herigault, J., Foggia, A.: Analysis of Shaft Voltages in Large Synchronous Generators. In: International Electric Machines and Drives Conference, Seattle, Washington, US, May 9-12 (1999)

References 187

[7.6] Link, P.J.: Minimizing Electric Bearing Currents in ASD Systems. IEEE Industry Application Magazine, 55–66 (July/August 1999)

[7.7] Cheng, S., Lipo, T., Fitzgerald, D.: Modeling of Motor Bearing Currents in PWM Inverter Drives. IEEE Trans. on Industry Applications 32(2) (March-April 1993)

[7.8] Hausberg, V., Seinsch, H.O.: Kapazitive Lagerspannungen und Ströme bei umrichtergespeisten Induktionsmaschinen. Electrical Engineering (82), 153–162 (2000)

[7.9] Hausberg, V., Seinsch, H.O.: Kapazitive Lagerspannungen und Ströme bei umrichtergespeisten Induktionsmaschinen. Electrical Engineering (82), 153–162 (2000)

[7.10] Drubel, O., Hobelsberger, M.: Medium frequency shaft voltages in large frequency converter driven electrical machines. Electrical Engineering (AfE) 89(1), 29–40 (2006)

[7.11] Amman, C.U.: Wellenspannungen in grossen, statisch erregten Turbogeneratoren, Dissertation ETH-Zürich (1988)

[7.12] Finkler, R., Unbehauen, R.: A new general equivalence transformation for mixed lumped and nonuniform distributed networks with synthesis applications. Electrical Engineering (AfE) 77(4) (May 1994)

[7.13] Simond, J.J.: Simsen Software (2004), http://Simsen.epfl.ch

[7.14] Lehn, P., Rittiger, J., Kulicke, B.: Comparison of the ATP version of the EMTP and the NETOMAC program for simulation of HVDC systems. IEEE Transactions on Power Delivery 10(4), 2048–2053 (1995)

[7.15] Drubel, O.: Current distribution within multi strand windings for electrical machines with frequency converter supply. Compel (International Journal for Computation and Mathematics in Electrical and Electronic Engineering) 22(4), 1166–1181 (2003)

[7.16] De Gersem, H., Mütze, A., Binder, A., Weiland, T.: Finite-Element Simulation of the Common-Mode Flux in Inverter-Fed Induction Machines, pp. 107–108. CEM, Aachen (2006)

[7.17] VDE 0141/5.76, Bestimmungen für Erdungen in Wechselstromanlagen für Nennspannung. VDE Verlag

[7.18] AIEE Committee Report, Voltage gradients through the ground under fault condition. Trans. AIEE Part III, 669–692 (1958)

[7.19] Drubel, O.: Basic Investigations of shaft-currents within screw compressors on account of residual magnetics. In: 3rd International Modelling School – Crimea 1999, Alushta, Ukraine, pp. 59–68 (September 1999)

Chapter 8

[8.1] Guardado, J.L., Cornick, K.J.: A Computer Model for Calculating Steep-Fronted Surge Distribution in Machine Windings. IEEE Trans. on Energy Conversion 4(1), 95–101 (1989)

[8.2] Wright, M.T., Yang, S.J., McLeay: General theory of fast-fronted in-terturn voltage distribution in electrical machine windings. Proc. IEE, Part B 130(4), 245–256 (1983)

[8.3] McLaren, P.G., Oraee, H.: Multiconductor transmission line model for the line end coil of large AC machines. Proc. IEE, Part B,132(3), 149–156 (1985)

188 References

[8.4] Manz, L.: Motor Insulation System Quality for IGBT Drives. IEEE Industry Applications Magazine, 51–55 (January/February 1997)

[8.5] Kaufhold, M.: Elektrisches Verhalten der Windungsisolierung von Niederspannungsmaschinen bei Speisung durch Pulsumrichter, Dissertation TU Dresden (1994)

[8.6] Espino-Cortes, F., Cherney, E.A., Jayaram, S.: Effectiveness of Stress Grading Cotings on Form Wound Stator Coil Groundwall Insulation Under Fast Rise Time Pulse Voltages. IEEE Trans. on Energy Conversion 20(4) (December 2005)

[8.9] Simonyi, K.: Theoretische Elektrotechnik, VEB Deutscher Verlag der Wissenschaften, Berlin, 5. Auflg (1973)

[8.10] Kupfmüller, K.: Einführung in die theoretische Elektrotechnik. Springer, Heidelberg (1965)

[8.11] Melfi, M., Sung, J., Bell, S., Skibinski, G.: Effect of Surge Voltage Risetime on the Insulation of Low-Voltage Machines Fed by PWM Converters. IEEE Transactions on Industry Applications 34(4), 766–774 (1998)

[8.12] Bell, S., Sung, J.: Will Your Motor Insulation Survive a New Adjustable-Frequency Drive? IEEE Transactions on Industry Applications 33(5), 1307–1311 (1997)

[8.13] Mbaye, A., Grigorescu, F., Lebey, T., Ai, B.: Existence of Partial Discharcges in Low-voltage Induction Machines supplied by PWM Drives. IEEE Transactions on Dielectrics and Electrical Insulation 3(4), 554–560 (1996)

[8.14] Bogh, D., Coffee, J., Stone, G., Custodio, J.: Partial-Discharge-Inception Testing on Low-Voltage Motors. IEEE Transactions on Industry Applications 42(1), 148–154 (2006)

[8.15] Bidan, P., Lebey, T., Neacsu, C.: Development of a New Off-line Test Procedure for Low Voltage Rotating Machines Fed by Adjustable Speed Drives (ASD). IEEE Transactions on Dielectrics and Electrical Insulation 3(4), 168–175 (2003)

[8.16] Lanier, C.: Using Corona Inception Voltage for Motor Evaluation. IEEE Industry Applications Magazine, 48–58 (1999)

[8.17] Kaufhold, M., Auinger, H., Berth, M., Speck, J., Eberhardt, M.: Electrical Stress and Failure Mechanism of the Winding Insulation in PWM-Inverter-fed Low-Voltage Machines. IEEE Transactions on Industrial Electronics 47(2), 396–402 (2000)

[8.18] Schäfer, K., Bauer, K., Kaufhold, M., Mäurer, A.: Insulation systems for high-voltage windings with optimized characteristics and the target to improve utilization for VSD-operation-capabilities and limits of conventional insulation. In: INSUCON International Electrical Insulation Conference (2002)

Chapter 9

[9.1] Bach, W.: Verlustleistung bei Umrichtern mit modernen Halbleiterbauelementen im Einsatz bei der Energieerzeugung. In: Proceeding of VDE-ETG Congress 2007, Karlsruhe (October 2007)

[9.2] Drubel, O.: Antriebstechnik in einem nachhaltigen Energieversorgungssystem. In: VDE Congress 2009, Düsseldorf (October 2009)

Index

A

additional losses 45
asymmetrical d.c. link 13

B

base frequency 158
boiler feed water pump 1, 155
bridge rectifiers 11
brushless exciter 1, 11

C

capacitive coupled shaft voltage 109, 118
cascaded-multi-level VSI 9
choke 113
circuit diagram 114
circulating eddy currents 47
circumferential flux 110
clamped multi-level 9
clamped multi-level VSI 9
coal fired power plants 1
coal fired steam power plant 154
constant motor power 158
constant torque 158
control strategy 22
converter applications 1, 153
copper losses 47
current path 119
current source inverter 9
cyclo-converter 9, 15

D

d.c. link 13
d.c. link voltage 16
damping 81
delay angle α 16

development directions 7
direct converters 14
double cage bar 49

E

eddy current distribution 46
eigen-frequencies 96
eigen-modes 95
eigen-vectors 81
electromagnetic force 86
end bells 52
energy efficient pump systems 153
energy industry 1
equivalent permeability 56
excitation system 111

F

Faraday's law 59
field weakening 158
five level cascaded converter 18
five level diode clamped converter 20
five-level converter 17
flux characteristic 17
forced commutating 16
friction losses 45

H

harmonic content 22
helmholtz equation 53
high frequency model 116
high speed drives 1

I

inductively coupled shaft voltage 112
industry segments 1
inertia moments 77, 79
insulation life time 141
insulation stress 135

190   Index

ionization 141
iron housings 52
iron losses 45, 50

L

lamination 50
load commutated current source
    converter 13

M

magnetic energy 72
magnetic force-density 92
magnetic stress tensor 92
massive magnetic material 52
material stress 84
matrix converter 22
maximum shaft voltage levels 130
medium voltage converters 17
multi-conductor arrangement 138
multi-level 18
multi-level converter 19
multi-mass-model 79
multi-phase diode bridge 9
multi-phase winding 12

N

naturally commutating 10
noise 85
numerical tools 7

P

partial discharge 141
piston air cylinder 97
pole surface 54
potential distribution 136
poynting-vector probe 63
pre-calculated pulse cycles 23
press-plates 52
pulse half time 143

Q

quasi stationary field problems 137

R

radial flux density 87
RC-module 111

Roebeled strands 48
Rogowski coils 119
rotating diodes 12
rotor losses 45

S

saturation factor 61
semi conductive layer 147
shaft grounding concepts 131
shaft voltages 109
single way rectifiers 11
six-pulse bridges 13
skin depth 55
skin time 115
space harmonics 69
space vector-modulation 23
starting device 110
stator d.c. copper loss 45
stiff motor design 1
stiffness 77
stiffness coefficients 79
stress grading coating 147
support rings 52
surge rise time 143
surge time 141
surge voltage 137

T

terminal voltage 17
thyristor bridges 113
torque oscillations 69
torsional shaft oscillations 77
transfer functions 99
turbo-generators 1
two winding layers 47
two-level bridge 16
two-level voltage source converter 16
two-level voltage source inverter 9

V

voltage peaks 137
voltage source inverter bridge 9

W

wave reflection 137
wind turbine applications 2

Printed by Books on Demand, Germany